food fashion and furnishings

Ann Burckhardt

Iowa State
University
Press,
Ames

Among numerous debts incurred in gathering material for this book, I owe a special one to **Robert Lindemeyer** for brotherly advice and a generous gift of time. I am also grateful to **Peggy Brown Black** for sympathetic help.

© 1984 The Iowa State University Press
All rights reserved

Printed by The Iowa State University Press
Ames, Iowa 50010

First edition, 1984

Library of Congress Cataloging in Publication Data

Burckhardt, Ann, 1933-
  Writing about food and families, fashion and furnishings.

  Includes index.
  1. Home economics—Authorship. I. Title.
TX147.B86   1984   808'.06664   83-10843
ISBN 0-8138-1941-5

The text in this book was printed from camera-ready copy supplied by the author.

Writing about food and families, fashion and furnishings

Writing about and families,

CONTENTS

Foreword by Lou Richardson and Genevieve Callahan — vii
Preface — ix

1. **Communications and you.** The challenge of sharing your expertise in writing and speaking — 3

2. **Your all-important audience.** The factors that distinguish the members of your audience — 5

3. **Getting started.** The five steps in the R/C formula for effective communication — 8

4. **Creative ideas, workable ideas.** The pursuit of strong ideas for writing — 13

5. **Writing readable copy.** The building of bridges for understanding — 17

6. **Leads.** The opening paragraph that no one can teach you how to write — 23

7. **The soft sell.** The subtle difference that service adds to selling — 28

8. **Photographs and other illustrations.** The integral part that visuals play in today's communications — 33

9. **Directions and how-tos.** The direct approach to writing how-to-do-its — 41

10. **Recipes and their adaptation.** The thinking that goes into a complete recipe — 46

11. **Meal plans and menus.** The supporting role that menus play — 55

12. **Features and columns.** The six types and how to do them — 60

13. **News releases and public relations.** The challenging tasks of the publicist — 69

CONTENTS

14. Magazine articles. The work that brings beginner's luck — 76

15. Booklets. The whats and whys of folders and booklets — 82

16. Cookbooks. The seven steps from idea to distribution — 87

17. Speeches. The writing that requires rehearsal — 94

18. Demonstrations. The four types and how to organize them — 98

19. Slide talks and filmstrips. The 10-step process from idea to screen — 104

20. Letters and newsletters. The requirements of these personal communications — 111

21. Copy editing and proofreading. The two final steps in writing for publication — 116

Afterword — 123
Index — 125

FOREWORD

How to Write for Homemakers was first conceived and written by us in 1949 for home economists who aspired to reach their audience with the written word. That the book filled a need is attested by the fact that since its first publication, it has appeared in five revised and updated printings of two editions.

Now, in the decade of the eighties, there is a need for its sequel. What is called for is an entirely new version tailored to today's fast-moving world, using today's language and today's visual techniques. And that is exactly what Ann Burckhardt has produced.

We are delighted with what she has done and we predict a great future for her book.

Lou Richardson
Genevieve Callahan

PREFACE

When I am interested in a job and I feel I'm qualified for it, I go ahead and take it. Not until afterward do I ask what the time demands will be and what rewards might be expected. Such was the case when the Iowa State University Press asked whether I would be interested in writing the successor to *How to Write for Homemakers* by Lou Richardson and Genevieve Callahan.

You bet I would. I had used this book during college and the early years of my career and had learned some valuable things from it. As I worked, moving from editing cookbooks to public relations to advertising consultant work and finally to newspaper feature writing, I learned much more. I learned it the hard way: by making wrong assumptions, by failing to consider all eventualities, by jumping to conclusions.

Not long before that call from Ames, I had begun working with fledgling writers. Like many a novice, these writers needed help with focusing a piece of writing, finding a lead, selecting detail and working out transitions.

Yes, I would try to sum up twenty-seven-plus years of brainstorms, interviews, rough drafts, rejection letters, rewrites, proofreading, by-lines and paychecks. It just might be fun. But distilling the wisdom that had been shared with me by a continuing series of managers and mentors would take some doing. To be exact, it took eighteen months of evenings and weekends.

What you have in your hand is really two books in one:

first, a primer on writing about your chosen field of endeavor; second, a how-to-do-it book on specific types of communications. This framework was retained from the previous book because it was so workable. Each chapter is complete in itself, with its own lead and as many examples as space allowed.

My aim was to construct a book that you will want to keep near at hand, so that you can reread a sequence here, rediscover a suggestion there. My hope is that it will smooth your passage from inexperience to expertise. I hope that afterward you'll be able to say, "I'm glad she told me that."

Ann Burckhardt

Writing about food and families, fashion and furnishings

1. Communications and you. The challenge of sharing your expertise in writing and speaking

WITH THE OPENING of this book you are seeking a new role: that of the communicator. You have come to realize that writing ability will widen your career horizons. Perhaps you have read one of the many articles and books predicting that communications will be one of the fastest-growing fields in the coming decades.

No matter what your area of expertise, you need to communicate effectively with your particular audience, whether its members are cooks, clients, consumers or colleagues.

The communication we are discussing in this book—writing about food, families, fashion and furnishings—focuses on useful ideas and information illustrated by both word-pictures and artwork.

Ideal communication, I believe, goes beyond presenting ideas and processes to interpreting what you know for those you hope to reach.

While some of your day-to-day communication will be oral—speeches, radio and television talks and presentations—a great deal of it will be written—letters, booklets, releases, even books. For some, writing comes as swiftly and easily as talking; for others, it requires long consideration and excruciating effort. In fact, an astute observer noted, "Very few people like to write, although most people like to 'have written.'"

Certain personal qualities seem to distinguish most successful writers. Those qualities are curiosity, fluency and judgment. *Curiosity* does not mean nosiness; it means your innate desire to know *who, what, why, where* and *how. Fluency* means an ease, a smoothness and a naturalness in your use of language. And *judgment* is that all-important ability to evaluate, to compare, then put first things first and get to the heart of the matter. Though some lucky writers are born with these qualities, all writers can cultivate them.

The indefinable extra that all truly able writers share is *empathy:* the ability to walk in another person's shoes, to see things through another's eyes, to feel another's hurt, or happiness.

Work with excitement

Writing, with its demands and its deadlines, requires dedication and self-discipline. But the dedication and discipline will never develop if you do not feel the excitement of taking a topic and developing it. When you write regularly you will discover how satisfying it is to take a bare-bones outline and flesh it out with facts and figures, anecdotes and examples. You already know how gratifying it is to suggest a new approach to someone on a one-to-one basis. In writing, you can multiply that good feeling many times over.

Writing for publication carries both a privilege and a responsibility. You, the communicator, have the privilege of sharing your expertise, of expressing your creativity, of advising others on how best to solve a problem. You also have the responsibility to present information honestly, accurately and fairly, and with all the clarity at your command.

Whatever your field, communication is sure to be one of its most challenging aspects. And developing communication skills will be most rewarding in any career you choose.

For further reading

Hellyer, Clement David. *Making Money with Words: A Guidebook for Writers.* Englewood Cliffs, N.J.: Prentice-Hall, 1981.

Mitchell, Joyce Slayton. *I Can Be Anything—A Career Book for Women,* 3d ed. New York: College Entrance Examination Board, 1982.

2. Your all-important audience. The factors that distinguish the members of your audience

YOU may never have stopped to count, but you are a member of many audiences. As a student, you act as an audience for your instructors. As a movie fan, you pay admission to join another audience. When you tune in the nightly newscast, you become part of a nationwide audience viewing worldwide events.

As a writer, you step out of the audience and onto the stage. Now, *you* are the featured speaker.

To visualize your audience, ask yourself: Who will read my article? Who will attend the new-product demonstration? Who will use the ideas in my booklet?

A decade or two ago, the answer was easy. Material about food, fashion, furnishings and families was aimed at Mrs. Homemaker, whose role was rather narrowly defined. But Mrs. Homemaker has been undergoing a major transition, and she is still gradually redefining her image. Nowadays, many homemakers pursue dual careers. Many are working wives. Some are single parents. Some are men. Daytimes they are busy in the workaday world; evenings and weekends they are equally busy in their homes and in the community.

The factor that unites members of your audience is the desire for a satisfying home life. Satisfaction at home takes many forms: rewarding relationships, nourishing meals, comfortable furniture, attractive clothing, festive get-togethers, even such basics as clean rooms and smoothly running appliances. The food, the equipment, the decor, the garments—all require the attention of the person responsible for running that home. If your writing can make home life easier or better, you are assured of an audience.

Please fill in: age, education, income

Members of your audience may differ widely in age, education and income. It's easy to see that a booklet titled *Teen Scene* would differ greatly from one called *After Retirement*. If your company has a research department, ask it for demographics (population studies). Or, check census reports for information about people in the areas in which your writing will circulate.

Materials of the type we are discussing will be used by men and women in the vast middle-income range. However, income levels and spending patterns vary greatly across this wide continent of ours. We Americans, once envied abroad for our affluence, may have more money in our billfolds, but it buys less. That's why its smart to study economics along with nutrition, family social science, interior design and fashion merchandising.

As you plan a communications project, you also will need some idea of your audience's education. While the majority of the members of your audience may not have attended school beyond high school, television has raised their sophistication greatly.

Consider, too, that many of the people in your audience live alone. Single persons now make up one out of five American family units. That figure is particularly important to food professionals developing recipes and meal plans. Although the trend toward solo living has been evident for several years, most recipes still are developed for four to six servings.

The amount of time your audience has for home activities and tasks should be kept in mind as well. When the homemaker is also a wage earner, the time factor may outweigh the economic factor. Your firsthand knowledge of what it takes to keep a home running smoothly is valuable. For example, the food writer who often eats out may not be able to weave the same helpful details into her articles as the writer who cooks daily.

Getting to know you

Many beginning writers of home-centered material get to know their audience by answering letters and phone calls that come into their offices. Communicating with people who are concerned enough to write or call gives the writer direct information about the needs and desires of his or her specific audience. Sometimes would-be writers can arrange student internships to get this type of experience.

Another good way to relate to your audience is to become acquainted with as many homemakers of varying ages and backgrounds as you can.

All the preceding factors about your audience—age, education, income, time—can be woven into a composite picture of your typical reader or listener. If you find it hard to address your writing to a thirty-one-year-old suburbanite with 1.3 children, try addressing an actual person—a friend, a relative, a neighbor—someone who closely matches that composite.

Remember, too, that you have two other audiences: your management and your colleagues.

The temptation, of course, is to consider yourself and your own situation as typical. Avoid that. As a college-educated person, you probably have more personal and financial resources than many members of your audience. In addition, you have studied home economics and are well acquainted with much of the basic information your audience may want and need.

For further reading

Elbow, Peter. "Section IV: Audience." In *Writing with Power*. Oxford, England: Oxford University Press, 1981.

3. Getting started.
The five steps in the R/C formula for effective communication

THE MAXIM "Well begun is half done" holds true for writing about fashion, food, furnishings and families. The more careful the preparations, the better the product.

Those preparations have been compressed into an easy-to-remember recipe by Genevieve Callahan and the late Lou Richardson, nationally known food consultants and magazine editors. Before retirement, the two served first as coeditors of *Better Homes & Gardens* and later of *Sunset* magazine. To help newcomers in home-centered writing, they summarized the steps they followed in their highly successful articles and books.

The R/C recipe for effective communications

Visualize
Analyze
Organize
Dramatize
Synchronize

See visions, dream dreams

First, *visualize your writing* in its finished form: the letter, the film script, the how-to-do-it manual. In some cases, your writing will be part of an established format—a news release

GETTING STARTED

or a monthly newsletter, for example. Then it will be easy to plan your copy so that you can make the important points within the space allotted.

But, more often than not, developing a format for a particular situation will require ingenuity. You will have to work out a variety of sizes and shapes in which the material might be presented, then pick the best format for your audience and your organization. This may require conferences with colleagues in printing, production and layout, plus a survey of publications similar to the one you are planning.

The ability to visualize what you are writing as it will appear in type and layout is an important asset. After you learn to do this, you will find it easier to write copy that fits the space. Experience helps here.

You're in analysis

Second, *analyze your problems*. Except for an occasional dream project (one that goes so easily that everyone is amazed), creative projects have a way of hitting snags. A key person is not available for an interview and another must be found on short notice. Or, the hall you planned to use for a program has no sink, so the demonstration has to be reworked so no running water is needed.

In a project assigned to you, never hesitate to ask questions about all the eventualities involved. The sooner you know what the problems might be, the sooner you can work out solutions. With experience, you will discover that projects with the most problems initially often bring you the most compliments. A problem-laden project forces you to bring to bear every bit of ingenuity. In so doing, you invest a lot of thought and generate a lot of enthusiasm.

Let's get organized

Third, *organize your thinking*. If a piece of writing flows right along, one idea developing logically from another, the writer probably used an outline. In grade school, we all grumbled about having to do outlines, yet outlines are alive and well. That's because they work.

Preparing an outline makes sense because it forces you to think through your project. And it helps you to recognize extraneous ideas. It guarantees a beginning, a middle and an ending, elements that readers expect. A word or phrase outline probably will suffice. That's assuming you know your subject thoroughly and have at hand lots of details for elaboration.

The first step in organizing with an outline is deciding on your purpose—the theme or idea you want to present. Let's say you are writing about the importance of a basic suit in a woman's wardrobe and have tentatively titled the piece, "Suit-ability."

Your research shows that selecting an important garment such as a suit involves three main considerations: fabric, color and style. Three is a frequently used number of subtopics within an article, but to develop your idea don't hesitate to use two, four or even more if necessary.

Some topics are easier than others to define—and confine. Using an outline will prevent you from going off on a tangent. For example, when writing about different types of lettuce, you might be tempted to move on to ideas for a salad bar. But that's an excellent topic on its own, so save it for a separate article or column.

All drama's not on stage

Fourth, *dramatize your presentation.* Most home-centered articles can benefit from dramatization. It makes an idea vivid or striking and therefore easier to remember.

Striking illustrations, either photographs or drawings, often are used to dramatize a piece of writing. Many editors use illustrations simply to grab attention. But in writing about fashion, food, families and furnishings the right illustrations carry vital information and save valuable space. Take, for example, a booklet detailing a simplified method for embroidering a blouse. A clear, close-up photograph or line drawing can show how to do it and attract interested eyes at the same time. (See Chapter 8 for more on the use of photographs and artwork.)

GETTING STARTED

With or without illustrations, you can dramatize your writing with *color words*, the apt adjectives, adverbs and active verbs, which add vitality and variety to all types of writing. For examples of words that conjure up imagery, scan the advertisements in a popular magazine or the commentary in a national news magazine. Or listen to the color announcer on a sports telecast.

Scenarios and personification are other devices for dramatizing ideas so that readers will remember them. Taking a cue from filmmakers, many writers and publicists now use scenarios. A *scenario* is a depiction of one or more situations, often rich with descriptive detail and mood, through which the topic or product is introduced.

A famous *personification* is Betty Crocker of General Mills, Inc. Back in 1926, the company's management decided that answers to homemakers' inquiries should be signed by a personable but fictitious woman, rather than by the impersonal home service department.

Let's synchronize our schedules

Finally, *synchronize your writing project.* The original use of *synchronize* had to do with the timing of machinery, but it has come to mean a merging or meshing together of various elements.

From conception to completion, your work on your communications project will have to be synchronized with that of others on your production team. After you visualize the color scheme for your booklet, you will want to synchronize that aspect with the purchasing agent who orders paper and ink. As you plan the illustrations for your feature article on home remodeling, you will want to inform the photo department that you need "before" and "after" pictures.

Communication among departments is vital to staying "in sync." If the advertising department knows that your bread-baking pamphlet will be ready October 1, it may be able to offer the pamphlet as part of an advertisement for yeast in the October issues of homemaking magazines. If the publicity department knows well in advance that wine red is the "in" color for fall, it can emphasize that shade by printing its advertising fliers and in-store posters in that same rich, red hue.

Synchronization may require some flexibility and adaptability. Suppose your new instruction manual is all laid out when your printer decides to install a new press that will permit a different, handier page size. After you talk it over with your co-workers, you decide that the new size would be better after all, so you regroup, reschedule and redo.

For further reading

Tarshis, Barry. *How to Write Like a Pro.* New York: New American Library, 1982.

4. Creative ideas, workable ideas. The pursuit of strong ideas for writing

IDEAS for writing about families, furnishings, food and fashion are all around you everywhere. Deciding which idea to use for a particular piece of writing is what causes concern for both writers and their editors or managers.

Ask yourself: What angle will put this information in a new light? Which approach will provide a framework for all the illustrations and examples I have gathered?

Successful ideas for nonfiction writing must be both creative and workable. No, these elements don't oppose each other. Effective ideas must have both the spark of imagination and the strength of practicality.

Successful ideas, experienced writers attest, are firmly rooted in the piece's purpose. Call it your objective, as the educators do, or your goal, as the planners do, but think about it long and hard. What do you want your audience to do? Do you have something to sell, or are you trying to change attitudes?

Often your second idea is much better than your first; the first was a mental warm-up.

Think purpose-fully

Let's look at two beginning writers and how their knowledge of their purposes influenced their work.

A home economist with a dairy co-op was asked to write a feature article for the members' magazine. She had a dual purpose: serving the readers—primarily the members' wives—and promoting the foods produced by the co-op. After learning that the women often joined their husbands in the field, she focused the article on a complete take-to-the-field meal that could be made ahead of time. The foods, of course, were rich with cheese, butter and even ice cream.

A homemaker who wrote free-lance articles wanted to sell a piece to a new magazine aimed at working mothers. With saving time as her purpose, she chose dovetailing household duties as her topic. Her source was material from a household equipment course. She updated those ideas, added more from her own experience and titled her outline "Two Tasks for the Time of One."

Two for the thought of one

Superiors often give beginning writers the general idea for a piece. If that is the case, your flair and originality can be brought to bear in the selection of information to include and in the writing itself.

Frequently such assignments are bare-bones subjects. For example, a newspaper staff writer was assigned a feature on onions. What *about* onions? There were many possibilities: how to select the best onions; how to cut onions without crying; how to combine onions with meats; how to prepare onion salads and relishes.

After scanning the information on onions in a food encyclopedia and the publicity releases on onions in the newspaper's files, she suggested two approaches. The first,"Onions and Other Lilies," would discuss the entire onion family, from chives to leeks. The second, "Onion Power," would be a tongue-in-cheek proposal for a new movement dedicated to the onion and its strength as a flavoring and accompaniment. Both pieces, it was understood, would include onion-rich recipes.

The presentation of two ideas is simple self-protection. If you suggest just one angle and it is rejected, you have to scramble for another. So try two or more. One surely will get the nod.

Wanted alive: a big idea

As your abilities and interest in writing develop, you will come to a point where you are seeking a *Big Idea:* an

CREATIVE IDEAS, WORKABLE IDEAS

attention-getting, promotable idea, an idea that will grab your audience. A Big Idea is, for example, a catch phrase that could be the theme for a national ad campaign, or a product name that would start a publicity blitz. Big ideas generate hundreds of words, dozens of pictures.

For a variety of ideas from which to choose, consider a *brainstorming session,* a session in which a group of people get together and try to solve a problem by spontaneously suggesting solutions. Everyone chimes in while one member takes notes. Negative reactions are not allowed—they stop the flow of ideas. *Piggybacking,* taking part of another person's idea and riding (or building) on it, is encouraged. Once you've brainstormed, you know it's fun, often exhilarating, and productive. A group of four or five can come up with a long list of ideas in a short time.

Most of the time, however, you must brainstorm with yourself. I call it "noodling," searching your "noodle" for a fresh look at a subject.

Free association is another way to work toward a creative idea. In this technique, borrowed from psychology, you allow your mind to freely associate one thought with the next. Use pencil and paper. And don't be surprised if you produce a chuckle along the way.

For example, take pies. A free association might go something like this: pie; apple pie; mom; warm; sweet; spicy; rolling pin; cut-outs; crisp; flaky; golden crust; the upper crust. That's it! The upper crust, the rich, the beautiful people, the first families. You could discuss the finest pies. Or, you could take the upper crust idea more literally and present a variety of pie toppings, such as meringue and streusel.

The intriguing thing about idea development is that two people will rarely suggest the same theme.

Dear notebook

Idea people frequently are either prolific note takers or faithful journal keepers. So, begin jotting notes on what you see, what you overhear, even what you eat. Often something that you notice piques your curiosity, but you don't have time right then to ask or read about it. When that happens, record that idea. Chances are, if you're interested in it, readers might be too.

Idea people also train themselves to spot trends. The fashion world is built on trends, but all other areas in home economics show trends, too. Trends, once begun by the daring and the innovative, may be seized and expanded with amazing speed. The writer who can get the jump on a trend and learn all its "whys" and "wherefores" can count on many story angles.

Timeliness and timing strongly affect the selection of topics for writing. For example, suggestions for saving money on household purchases may be welcome year-round but are especially appreciated in the early months of a new year, when Christmas bills and taxes must be paid.

Fledgling writers often worry that their topics are not original. Good ideas in homemaking are seldom new, in the sense that they have never been presented. More often, the idea you develop will be an adaptation or a modification of an old standby. And don't let a critic deter you by saying that your ideas are derivative. Good ideas never die; they just reappear in new contexts.

You will surely be most enthusiastic about ideas that you like and feel are worthy. But when your supervisor picks another, perhaps unfamiliar, idea, delve into it. Read all you can, interview experts. The interest you generate as you pursue the topic is bound to be transmitted to your readers.

For further reading

LeBoeuf, Michael. *Imagineering, How to Profit from Your Creative Powers.* New York: McGraw-Hill Book Co., 1980.

5. Writing readable copy. The building of bridges for understanding

"MOMMY has a lot of words in her fingers," the toddler reported to her grandmother after weeks of playing in the study as her mother worked as a free-lance writer.

The child might have been mistaken about the source of the words, but she already had experienced the fascination that words can hold.

Words are to the writer what steel is to the builder and colors to the artist. Like the builder, the writer must construct carefully, and, like the painter, choose consciously.

The analogy of the builder is particularly appropriate for the writer specializing in food, fashion, families or furnishings. As writers, our aim is to build bridges between the experts and our homemaker/consumer readers. To ease our reader's travel, those bridges must pass inspection for accuracy, brevity and clarity.

Be resource-full

Before we get into the important qualities of good writing, we should survey the tools and training you need. The fiction writer or poet may need only paper and pencil, but writers who are specialists require additional resources, such as reference materials and professional friends and experts.

Your reference materials will include important books in your field, such as texts on the history of food, furniture or fashion. They also will include source books and professional journals from a general or technical library.

And, added to your own experiences and observations, your personal resources will include those of your network—

the social and professional friends with whom you trade counsel and encouragement. Some of them may be experts in your field whom you can interview by telephone, by letter or in person.

Go into training

Your training as a writer actually began whenever you began to read. Most writers are avid readers and surround themselves with magazines and books. Once you begin writing you become a more discerning reader. Whether you are scanning the latest fashion magazine or poring over technical journals, you are bound to spot a fact or figure, an anecdote, or even a pattern of organization that you can use in the future. The developing writer should consider time and money spent on reading and reading materials a worthwhile investment. What's more, reading is a fine form of relaxation.

Your grade-school English lessons, from declensions to diagramming, **are** another key element in your training. Correct grammar and perfect spelling are essential trademarks of the professional writer.

As with learning to play the piano, learning to write requires practice; daily practice, if possible. So, beginning today, reread your reports, essay questions and even your personal letters to check them for such common errors as marathon sentences, changes of tense and pronouns without antecedents. Developing writers need feedback, too. If possible, seek the counsel of an experienced writer who will work with you on your rough drafts. Comments from such an adviser are good preparation for the critiques from your editor/manager.

If you **are** a student, take your notebook along on semester breaks and try your hand at a haiku or a slice-of-life short story, just to keep in practice. Or, better yet, scout out a feature to free-lance while you're away. If you are working at an entry-level job, set aside an evening for a writer's group and share your efforts with others, trading comments and criticisms. If you are a homemaker, volunteer as newsletter

WRITING READABLE COPY

editor or publicist for your parent-teachers association, your church or a nearby hospital.

Another hint: plan to write at the high point of your day in terms of energy and alertness.

And, when working on a long writing project, take a break when the words are flowing easily rather than when you have come to the end of a section. That way, it's easier to pick up the thread.

Now, let's discuss the writer's ABCs—accuracy, brevity and clarity—in more detail.

Accuracy means hitting the mark

Strictly speaking, accuracy means that you verify names and titles, that you quote your interviewee correctly and that you describe the steps in a process precisely. Broadly speaking, accuracy means meeting a high standard of performance that begins with faithfulness to your sources and ends with giving your readers all they need to know.

Checking and double-checking your copy to be sure that every fact and figure is right is time-consuming and boring, but it's absolutely necessary. Editors, you'll find, consider an entire manuscript suspect if they discover one inaccuracy. And, in some organizations, copy is reviewed by a series of executives, each of whom may spot a dubious angle or an incorrect implication.

Aim for accuracy in your choice of words. If the designer labeled the new suede shoes "root beer" color, use that rather than reddish-brown. Fine-tune your writing by using exact terminology. For example, the cook uses water to "reconstitute"—not mix—frozen orange juice concentrate.

It is infinitely more interesting—as well as more accurate—to write, "the waitress *slid* the lutefisk onto the diner's plate" than to write "she placed the slippery Scandinavian specialty on the plate." Instead of describing a man as "tired," note that he is "slumped in posture and drawn of face."

When in doubt, use a simpler word in place of a more complicated one, use "decide," not "finalize"; use "hope," not "expectation"; use "use," not "utilize."

The ultimate accuracy test lies with the reader. When writing for others trained in food science, we may discuss the "gelatinization of the starch," but for homemakers say, "Cook the flour mixture until thickened."

Brevity for speed and ease

Today's readers and listeners are busier than ever. They want to get to the heart of the matter. Therefore, keep your writing brief and to the point. Pin this motto above your desk: "Write it tight."

Brevity can be achieved in many ways. One is to write an outline of your brochure or article first, then stick to it. The outline approach takes discipline—self-discipline.

To write with brevity, write simply. It's sometimes tempting to include all the options. For the sake of brevity, it may be necessary to skip the fact that mushrooms or nuts can be added to the recipe.

Use active rather than passive verbs. Recast "The gourmet dinner was prepared by the chef." to "The chef prepared the gourmet dinner." Or, for variety's sake, substitute "concocted" for "prepared."

Many writers, and many more editors, cut copy to achieve brevity. Sounds ruthless, doesn't it? Ruthless but right. Introductory material often can be slashed. You, the writer, needed to warm to the subject, but the reader, already attracted by the title or illustration, is ready for the information. Frequently, an entire piece of writing needs trimming or tightening. Try to do the first draft of your article well ahead of deadline, then sleep on it. Later, cut it substantially. When you reread that second version, you'll probably find that a pet phrase or two had to be deleted. Short, to-the-point writing is more valuable to today's reader than clever couplings of words.

Every sentence should carry its own weight. Every paragraph should be necessary. If you have made your point, there's no need to restate it.

And the greatest of these is clarity

Of the writer's ABCs, clarity is the most vital. All roads lead to clarity: focus, transitions, organization, choice of words, everything.

When you focus your writing, you give it sharp detail: no fuzzy thinking, no generalized descriptions, no unconscious repetition. Like the photographer, you decide what point of view is best for the word-picture you're creating, then stick to it. If you begin a piece on trends in window treatments from the homemaker's view and shift midway to the drapery manufacturer's viewpoint, your article will be a confusing blur.

The writing of transitions is one of the hardest techniques for neophyte writers to master. To be effective these connectives—sometimes only a word or two, sometimes an entire anecdote—must be smooth and natural. For practice, stop reading right here and pick up another publication, perhaps today's newspaper or a favorite magazine. Using a marking pen or crayon, color over the transitions in an article or two (it isn't necessary to read the entire piece). Then, scan the article without the transitions. Stilted, isn't it?

One practical editor calls transitions the glue that holds the article together.

But even clever transitions can't rescue a piece of writing that is sloppily organized. Jot down your plan of organization before you begin writing. Or better yet, work out two ways of organizing the material and pick the better one.

Many how-to-do-it home articles require start-to-finish organization. Some feature articles readily lend themselves to chronological order, or, if the subject is interesting enough, to a flashback treatment. For example, an article about a new product often opens with its uses, then moves to a detailed description, and, finally, to its development.

Occasionally, the need for clarity requires rewriting. Sometimes, too, your editor or manager decides an article should have a different focus or a different length (usually shorter). As you get busy and redo your piece, remember that many novelists rework a single chapter four or five times.

The last word on writing: meet your deadline.

CHAPTER 5
For further reading

Houp, Kenneth W., and Thomas E. Pearsall. *Reporting Technical Information,* 4th ed. Encino, Calif.: Glencoe, 1980.

Strunk, William and E. B. White. *Elements of Style.* New York: Macmillan, 1972.

The Writer, monthly, published in Boston.

Writer's Digest, monthly, published in Cincinnati.

Zinsser, William. *On Writing Well,* 2d ed. New York: Harper and Row, 1980.

6. Leads. The opening paragraph that no one can teach you how to write

AS CHILDREN, we played follow the leader, merrily mimicking movements. As readers, we follow the lead of the writer, reading only so long as the topic holds our interest. As writers, we lead the reader into a piece with an opening statement called, not surprisingly, a lead.

The *lead* (sometimes spelled *lede*) is a concise yet catchy paragraph that puts the subject in a nutshell, so to speak.

Some writers are so sure of their approach that they work out their lead even before they begin gathering information for a piece. Others mull their material for days until one apt phrase comes to mind.

When well conceived, your lead becomes the foundation for your article or press release. And restatements of the lead can provide effective transitions within a piece.

No teacher, no textbook can tell you how to write your lead. You must learn by doing it.

You also can study the leads in your favorite magazines and newspapers. Don't try to read the entire article. Simply scan the opening sentences as quickly as you can. Certain patterns will emerge as you compare other writers' leads. Many include a surprising amount of information.

A good news lead usually includes at least four of the famous "Five W's:" *who, what, why, when* and *where*. It's the *why* that's difficult to squeeze inside that nutshell.

A good feature lead may concentrate solely on the *what*—the topic.

The sample leads below are grouped according to the literary devices their writers have used.

Contrast leads

Remember all those "compare and contrast" essay questions in school? The lead writer can compare present with past, here with there, familiar with unfamiliar, the way things seem with the way things are.

> It used to be that the scholarly approach to fashion, wherever expressed, was a backward glance...study and presentation of costumes worn by leaders of ancient civilizations and princesses of primitive cultures. No more. Today, the "ivory tower" view of fashion is expanding to include the artfulness of modern design and is inspiring nods of respect and gesture of commitment from unexpected precincts in the museum and education communities.
> *Vogue*

> The grape grower is a farmer by inclination but a prophet by necessity: His livelihood depends on the ability to predict, then plant, wine varieties that will please the fickle consumer.
> *Minneapolis Star*

Question leads

The well-constructed question lead carries the reader into the article to find the answer.

> If you can't afford diamonds and you can't stand rhinestones, what else is there?
> Not much, New York jewelry designer Marsha Breslow found. So she decided to do a line of good jewelry that would tie in with fashion trends and fill the gap between expensive investment jewels and cheap costume pieces.
> *Minneapolis Star*

LEADS

> No laundry room? How about a laundry closet? The compact washer/dryer and complete work area tuck into a double-door closet five feet wide and thirty inches deep.
>
> *Woman's Day*

Descriptive leads

Mood, color, spirit—all can be captured by the keen observer in a lead that invites continued reading.

> Such is the dry, airy radiance of the Athenian light and climate that one views the city's shortcomings—frenetic traffic and some ankle-twisting sidewalks—with indulgence: A kind of chaos is somehow all part of its infectious *joie de vivre*. Not everyone finds Athens as beautiful as its aficionados do, but nobody with a selective eye and an appreciative nose for the combined aromas of resin and incense, roasting corn, chestnuts and coffee, nobody with a zest for the contrasts that emerge from layers of civilization, could fail to find it fascinating.
>
> *Gourmet*

Staccato leads

As one-time piano students know, the staccato lends excitement to music, and so does a rapid-fire staccato lead.

> Brennan's. Antoine's. Bourbon Street. Gumbo and shrimp Creole. Late-night jazz, early morning *cafe au lait*. Shrimp, oysters, crayfish and crab...these mean New Orleans to many Minnesotans.
>
> *Minneapolis Star*

Figurative leads

The use of figures of speech, including metaphors and similes, requires a combination of flair and care.

> The year of 10.8 percent unemployment has now roosted on the much-dampened knee of Santa Claus, who came into town already harassed by a corps of psychologists that thinks he is a needless social antique.
> *Minneapolis Star & Tribune*

Parody leads

The parody, an imitation undertaken for effect, comic or otherwise, is the frequent ploy of the lead writer with a literary bent. Song titles, famous sayings and names of best sellers and box-office hits are popular sources of parodies.

> Guess what's coming to dinner—or breakfast or lunch? Your favorite houseplants. And the idea is as practical as it is decorative. By using plants as centerpieces or favors, you save the expense and bother of buying perishable cut flowers.
> *Better Homes & Gardens*

Word play leads

The devices of the poet and the punster can be applied to your leads, often with ease and, occasionally, with telling effect.

> Dagwood would be delirious.
> The indecisive diner would be dumbfounded.
> All because of the Brine Meat Market's offer of over 44,000 varieties of sandwiches, made to order, hot and cold, on dark or light, with or without trimmings.
> *Minneapolis Star*

Quotation leads

When a feature article focuses on a personality, a quote from that person is an obvious choice for the lead.

> "I've always had fun," said Stanley Marcus, chairman emeritus of Neiman-Marcus, over a telephone from the back seat of his limousine, which was whisking him across his home town of Dallas. "I think that's why I've always liked gifts that have a touch of whimsy; gifts that make people laugh."
> *Minneapolis Star & Tribune*

Anecdote leads

Short vignettes or anecdotes frequently are used in magazine leads. Their drawback is that they do little more than introduce the topic and the writer's approach to it.

The lead on Chapter 5 is an anecdote lead.

Unconventional leads

Some writers have a knack for writing simple, almost casual, leads that launch a piece of writing in a low-key, conversational manner. In fact, one editor terms this approach "the falling-off-a-log lead."

> "For those who belong, the private club is like home."
> *Minneapolis Star*

> "A T-shirt is a sort of a human bumper sticker."
> *Minneapolis Star & Tribune*

If lead writing seems difficult at first, remember that leads pop into our minds when we least expect them: while doing housework or driving. The lead comes from our subconscious—some time *after* our consciousness has been agonizing over it.

For further reading

Emerson, Connie. "Great Beginnings and Happy Endings." In *Write on Target*. Cincinnati, Ohio: Writer's Digest Books, 1981.

7. The soft sell. The subtle difference that service adds to selling

BUYING AND SELLING make the business world go round. If you are—or expect to be—working in the world of business, you'll need to know how to sell with the written word.

If you think the word "sell" is a bit strong, try "persuade." But, in a very real sense, a menu writer sells entrees to the restaurant's customers. An extension agent sells the latest, most efficient canning method in her weekly newspaper column. A food company sells a home economics class an unusual party idea via a brochure. And a fledgling copywriter sells a publicity campaign theme to the boss.

I've used the term *soft sell* because the type of sale we're talking about is invariably enhanced by its uses, benefits and/or problems solved. The *soft sell* is subtle, service-oriented, often subliminal, while the *hard sell* is direct, money-oriented, often demanding.

For example, one dishwashing detergent builds its advertising campaign on a lovely liquid that yields both beautiful hands and clean dishes. Its competitor concentrates on value/cost comparisons—you can wash more dishes for the money.

You, the salesperson

As a specialist in fashion, food, families or furnishings, you have three advantages to offer on behalf of your business, school or community organization: ideas, skills and judgment. You keep up-to-date by watching trends and noting what's new in your field. You maintain a file of ideas that are valuable from the standpoint of both the homemaker/consumer—and your organization. What's more, you continue to grow professionally as you work. You're learning on the job as well as in classes and seminars.

While reliable judgment requires experience and a little luck, you shouldn't hesitate to voice your opinions. Your experience, your professional contacts, plus a fresh viewpoint, lend validity to your judgment.

And don't be afraid to trust *intuition*, that elusive gift of reading moods and emotions. It often informs judgment.

You make the team

Each person has different skills, ideas and experiences. That's why many of us work as a member of a team. It stands to reason that by working together, teammates can avoid duplication, divide responsibilities and accomplish much more than they could alone.

Many challenging careers are offered in advertising. You may start in the consumer service department, interpreting consumers' needs to your company's advertising staff. You may work in the creative department of the company's advertising department or in an advertising agency. Or you may serve as a free-lance consultant to a company or agency.

Wherever you work, try to remember these sure-sell advertising approaches:

- promise a benefit and prove it
- solve a problem
- present a good idea
- show how to use the item

Saving time is an easy-to-prove benefit when you plan advertising for items used by the family in the home. This is true for a wide range of items, from nonstick skillets to pleater tape.

Solving problems is the bread and butter of television advertising for home products. You'll see a half-dozen swift scenarios in an hour of television viewing. The child spills food on the carpet and mom solves the problem with a can of "cleaner-upper." Or the cook comes home late and solves the supper problem with a frozen entree.

In the good-idea department, the possibilities are endless. Given a new terry towel, a creative person can suggest using it for anything from a sarong to a placemat.

For years, a national soup manufacturer has built its sales campaigns on how-to-use-it ideas. One year it pushed saucery—the soup as a sauce; another year it was soup-mates—how to mix two soups to create a new flavor.

From inspiration to imprint

To illustrate the soft sell, let's trace the creation of a food advertisement.

Theme: the basic approach of an upcoming ad campaign is established by the client company and the advertising agency. A series of ads is to be created for magazines, newspapers and television.

Inspiration: a layout artist has an inspiration for an illustration. S/he works up a rough layout and submits it to the account executives, who deal directly with the client company.

Implementation: the account executives think the idea worth pursuing and call in the food professional to suggest foods that carry out the theme. S/he offers several possibilities, then develops recipes that could be features in the ads.

Claims: if the copy accompanying the advertisement makes claims about the food, such as health or energy-saving benefits, the legal department must have time to substantiate those benefits. Government agencies regulate such claims strictly.

Goal: as far as the agency executives are concerned, the idea is what counts. It probably will come from the copywriter, but it could come from the food expert or the artist. All three must work closely, brainstorming, exchanging thoughts and modifying the basic idea.

Everyone involved in a new food advertisement tries for an idea that is practical yet has that indefinable something extra. Call it excitement or originality. It stops the casual

THE SOFT SELL

readers, causes them to look twice, to read the headline and copy and to consider the product favorably. Unless the advertisement results in a sale, it has not achieved its goal.

It may seem farfetched, but selection of a recipe for a multimedia advertising campaign is much like judging a beauty contest. The subject must be photogenic, appealing, representative and versatile.

When it is up to you to pick the best recipe for a new campaign, you might ask yourself:

- Will the food be visually exciting in the photograph? Can it be made the center of attention? If it isn't colorful, can it be garnished to look appealing? Can the photograph register the quality of the product, such as the succulence of the ham or the moistness of the cake?
- Does the recipe use enough of the product—certainly more than a quarter of a cup? Does the recipe's success depend on the product? Does it demonstrate the quality of the product?
- Is the recipe up-to-date? Will the ingredients be readily available in the area where the ad will appear? Are the directions clear and easy to follow?
- Does the dish have potential for long-standing popularity? Will it become the sort of recipe that is passed along from cook to cook? Does it lend itself to variations?
- Does the recipe have an appealing name? Don't be surprised if the copywriter tinkers with that. Remember: A good recipe name is a headline in itself.

Writing your own advertising

As your career progresses, you may decide you'd like to be your own boss, to become an entrepreneur. Perhaps it will be a neighborhood shop or a consultant service; perhaps you will teach your special skills from your home. If you do begin your own business, one of your top priorities will be advertising.

First, you will want to write an announcement letter (see Chapter 20 on letters and newsletters) to professional colleagues and potential clients. Get as many eyes to read that letter as possible before you send it to the printer, so that it's free of typographical and grammatical errors.

Next comes a brochure (with your photograph) describing your services, hours and fees. Here you will expand the material for your résumé, putting your best foot forward.

Before long, you may want to run a small advertisement in a neighborhood publication or a professional journal. Sketch an idea, then take it to someone who specializes in ad copy. The advice of an experienced advertising writer can help you avoid obvious pitfalls, for example, the overuse of professional jargon.

For further reading

Advertising Age, the international newspaper of marketing, weekly. published in Chicago.

Johnson, J. Douglas. *Advertising Today.* Chicago: Science Research Associates, 1978.

Ogilvy, David. *Ogilvy on Advertising.* New York: Crown Publishers, Inc. 1983.

8. Photographs and other illustrations.
The integral part that visuals play in today's communications

WHEN we were very young, we delighted in "reading" our picture books, studying the illustrations and putting together a story. As adults, we still do much the same thing—scanning the photographs or illustrations before reading the text.

Artwork is an integral—and increasingly important—part of most communications about food, fashion, furnishings and families.

Why? Because the reader is highly sophisticated when it comes to visual imagery. Thanks to television, Americans have attained a high level of visual literacy; we readily read the meanings in pictures. Certain elements in photographs tell us certain things. A measuring tape says the foods pictured are diet foods; a Christmas ornament in a publicity photograph shows that the garment draped over the box is to be a holiday gift.

Types of photographs

The photos that you, your editor or manager, the art director and/or the photographer select to illustrate your writing will be one of five types.

First, there's the *stopper:* something bold to catch and hold attention.

Then there's the *space saver* or *amplifier:* the picture that saves words (but perhaps not 10,000!).

Third on the list is the *storyteller:* pictures that can speak louder than words about a situation, a solution, even a viewpoint.

Fourth is the *sequence:* How-to-do-it pictures, sets of before and after shots, an album showing how a person grows and changes; there are many examples.

And last, comes what I call the *eye saver:* a decorative element, usually small, that offers visual relief in a layout.

To assess the state of the art of illustration, stop here and look at a sample of publications near-at-hand. Look through a textbook, a news magazine, a home-living magazine, a fashion magazine and a newspaper. Study the articles and the advertisements. Try to find at least one example of each of the types of illustrations discussed. If you find several of one type, ask yourself which is the most effective. Can you explain why?

After studying a variety of photographs, you know that one well-thought-out photo can serve more than one purpose. For example, an actual-size picture of a 6-inch panorama egg (a sugar shell with a scene inside) acts both as a stopper and as its own storyteller.

When are they used?

Which of the five types of illustrations are most frequently used with copy on families, furnishings, fashion and food? Both the storyteller and the sequence appear often. Each of these two types can, in turn, be broken down into combinations.

A storyteller can show a product: with its creator; in a setting for appropriate use; or with a still life of its components. For example, a prize-winning chicken recipe might be pictured with the cook who concocted it, *or*, on a buffet table with suitable accompaniments, *or* on a counter surrounded by its ingredients. Newspaper and life style sections usually print the product-plus-creator photo. Publicists select either the product in its setting or the product with its components—and sometimes both.

English Pub Cooking With Step-By-Step Techniques

Meat mixture is spooned into center of pastry and edges of pastry are brushed with egg before sealing.

Seal the edges of the pastries with a fork. They are now ready to be baked to golden-brown.

Serve Dickens Pastries pub style with a tankard of ale and a boat of judiciously seasoned gravy on the side.

Venture into any pub in England at lunchtime and chances are you'll be served "pasties" with your tankard of ale. pasties are traditional English-style "sandwiches" usually filled with cooked beef, pork, fish or lamb. Here, Franco-American Beef Gravy, in the pasties and the flavorful sauce, adds a full, rich taste and gives body.

The parsimonious English are champs at using leftovers to best advantage. Dickens Pasties are proof positive that a pittance of beef with additional ingredients can be stretched to serve six.

DICKENS PASTIES

1½ cups finely chopped cooked beef
1 can (10¼ ounces) Franco-American Beef Gravy
2 tablespoons finely chopped onion
1 teaspoon Worcestershire
1 teaspoon prepared horseradish
1 package (10 ounces) pie crust mix
1 egg, well beaten
2 tablespoons chopped pimiento
¼ cup chopped parsley

STEP 1. Mix *thoroughly* beef, ¼ cup gravy, onion, Worcestershire, and horseradish.

STEP 2. Prepare pie crust as directed on package. Roll out pastry; cut into twelve 5-inch rounds.

STEP 3. Spoon about ¼ cup meat mixture in center of each of 6 rounds. Bruch edges with egg; top with remaining pastry. Seal edges with fork. Slit tops; brush with remaining egg.

STEP 4. Place on baking sheet; bake at 500°F for 10 minutes or until brown.

STEP 5. Meanwhile, in saucepan, combine remaining gravy, pimiento, and parsley. Heat; stir occasionally. Serve with pasties. Makes 6 servings.

(Courtesy Campbell Soup Company)

Storyteller-sequence photograph

A well-done food photograph is three ideas rolled into one—the recipe, an attractive serving idea, and a complementary menu.

Sequence art, whether photographs or sketches, takes the typical forms. We have come to take step-by-step sequences for carrying out tasks for granted. Magazines use them regularly to show techniques. Home products (zippers, for one) often enclose step-by-step art to help the buyer use the items.

"Before" and "after" sequence photos often are used in home decorating articles and in material on hairstyles and makeup. Depictions of alternates and coordinates are widely used, too. Models are often pictured wearing the same dress with various accessories. Or, the garments of a basic wardrobe are combined and recombined for different looks.

Color or black and white photographs

How exciting it would be if every photograph could be published in color. Thanks to color television, we take color for granted. But color transparencies and four-color plates are expensive.

Typically, black and white photographs are used for newspapers, newsletters and internal company publications, while color photos appear in magazines, filmstrips, booklets and brochures. In black and white, the key is contrast; in color, it may be either contrast or harmony. Unfortunately, a photographic setup that has been designed for color reproduction rarely photographs well in black and white.

Black and white photographs, of course, are not really black and white but various shades of gray. When you select the elements for a black and white setup, disregard hue and focus on grayness—light, medium and dark. Photographers and food stylists who work in black and white soon learn to narrow their eyes and judge the gray level of colored items being considered for a photo. A simple rule: Put dark objects on a light background and vice versa.

PHOTOGRAPHS AND OTHER ILLUSTRATIONS

The five W's of illustrations

When your job involves lining up illustrations, you'll want to answer the five W's, *who, what, why, when and where,* before the project gets too far along.

Why? Why use an illustration? Can't the copy carry its own weight? We use illustrations not only because readers expect them with their daily reading fare, but also because pictures contribute to content and impact.

Smart writers realize that copy set off by a compelling illustration attracts a higher readership than unillustrated material.

What? What type of illustration would be most effective: photo, sketch or graphic combining a photo and a drawing? Would the use of beginning-to-end pictures or sketches save a great deal of explanation? What information must the photograph convey? What items should be shown? Which of the six directions of photo composition would work best: *vertical, horizontal, diagonal, balanced, radiating or triangular?*

What is the purpose of the picture? Take, for example, a picture of a chocolate cake. If the purpose is to show texture, show a slice with the cut side facing the camera. If the purpose is to present a new birthday-party cake, get some party excitement into the picture. If the aim is to promote the use of walnuts, an action shot of a hand patting chopped walnuts around the side may be in order. If the cake is to grace a package, it must fit the format decreed by the package designer.

Who? When a photograph has been scheduled, take time for a conference—or conference phone call—to delegate duties before the legwork begins. Will you have the expertise of an art director, or will you or the photographer—or the two of you together—have to work out the photo's composition? Decide which person is responsible for the background, the foreground, the product and the *props* (properties, such as accessories and table settings). The food professional is nearly always responsible for the product, but frequently is expected to do the "propping," too. Some organizations, however, employ stylists who are responsible for the treatment the product is given; for example, a new housewares item may be shown in the context of a bridal shower.

When and Where? Good photographic setups, like good writing, demand a lot of thought and preparation. If you're rushed and not well prepared, you often end up further behind schedule because the photo turns out badly and has to be retaken.

Most photographers prefer to work with home products in studios, where lights and backgrounds can be arranged for best results. Occasionally, however, another location, either outdoors or in an evocative setting, is far better. For example, a period room at the local art museum would make a fine backdrop for an updated version of an historic food. The school playground, the swings or trapeze perhaps, is an obvious setting for a photo of back-to-school fashions.

Voices of experience

Learning to set up high-quality pictures requires on-the-job training.

For many years those working in food photography used special techniques to produce beautiful photographs. These techniques included substituting fondant for ice cream, topping desserts with swirls of white shaving cream and adding soap bubbles to coffee to make it look fresh. Some of these practices are still in use for magazine and newspaper photographs, but truth in advertising regulations have been enacted by the federal government and foods must be prepared for photography exactly as they are prepared for human consumption.

Here is an example of how a food company changed its photography practices to comply with federal regulations. In the past, they had placed marbles in the bottom of bowls of their soup which were to be pictured so that the solid ingredients would be near the top of the stock. Now, they have redesigned their photo layouts so that a spoonful of the soup is a key element. And they use very shallow soup plates carefully lighted so that the ingredients can easily be identified.

PHOTOGRAPHS AND OTHER ILLUSTRATIONS

These federal regulations are part of a national movement based on the consumer's right to know what is in the products they are purchasing. These regulations have in fact increased the work that must be done on food and product photographs because extra ice cream sundaes, souffles, etc. must be ready when the first ones begin to melt, deflate or otherwise react to the heat of the studio lights.

Here are suggestions from people who have spent long hours in photo sessions involving home products. But they are just suggestions, not rules; and for each suggestion, there's an exception. Ideas for food photography dominate the list because so much food photography is done.

- *Size:* If a larger or smaller photo than the typical 8-x-10-inch size is needed, or if a matte (textured) finish is required rather than a glossy, be sure the photographer knows.
- *Center of interest:* Only one item can be the photo's center of interest. Choose it carefully to be sure it's worthy of center stage, then arrange the other elements accordingly. When two foods are shown together, one must be dominant, the other subordinate.
- *Stand-ins.* Just like Hollywood stars, food subjects require stand-ins. Bake one cake to stand in while the props, lights and angles are arranged; have the perfect one ready to put in place just before the shutter clicks. If the food will wilt or melt or soften, you may need two stand-ins. And if you don't need one of them, the studio crew will certainly be glad to dig in.
- *Background:* Keep it unobtrusive, complementary.
- *Hands on:* Hands often are featured performing tasks. This technique allows a close-up of the process. If more of the person were pictured, the clothing and hairstyle might date the photo.
- *Simplicity:* It's the best policy. Stark simplicity bespeaks elegance. Ordinary, garden-variety simplicity speaks of accessibility; the viewer thinks: "I can do that."
- *Pattern:* Only one pattern per photograph, please. If the dishes are bordered, skip the patterned cloth or napkin. The same goes for fabrics in room settings.
- *Tableware:* Plain white dishes and unadorned glassware

are recommended because they do not compete with the textures and contrasts of the food. Remember the saying the Italians have: "The eye does half the eating."

- *Busyness*: Table-setting photographs get "busy" before you know it. A centerpiece, goblets, placecards, silver, salt and pepper shakers are not always necessary. Stick to the essentials that say what you want to say.
- *Garnishes*: The best garnishes are fashioned from one of the ingredients in the recipe or one of the accompaniment foods. An example is the triple-tier carrot cake decorated with carrots made of frosting, each clearly defining the shape of one serving and together forming a lovely circular pattern. And forget the parsley; it photographs black in black-and-white.
- *Preparedness:* Put together a little kit of handy items to take along when you're shooting on location, such as a pocket knife with a bottle opener attachment, moist towelettes, comb, pins (push, straight and safety), plastic tape, foil, whatever you think you might need. Don't forget a garbage bag and twist closures.

Sometimes, no matter how hard we work, the photograph is not as successful as we had hoped. That's the time to consider cropping and/or sizing it to redefine the center of interest. Ask an experienced layout artist or photo technician for pointers on sizing and cropping. If that fails, reshoot.

For further reading

Hurley, Gerald D., and Angus McDougall. *Visual Impact in Print.* Chicago: Visual Impact, 1971.

Rothstein, Arthur. *Photojournalism,* 4th ed. Garden City, N.Y.: AMPHOTO, 1979.

Salomon, Allyn. *Advertising Photography.* New York: American Photography Book Publishers, 1981.

Sauboa, Jean, and Associates. *Introduction to the Visual Arts.* New York: Tudor, 1968.

White, Jan. V. *Editing by Design.* New York: R. R. Bowker Co.

9. Directions and how-tos. The direct approach to writing how-to-do-its

THE UNITED STATES has become a nation of do-it-yourselfers. From carpet laying to candy making, we do it ourselves to attain a higher standard of living, to save money, even to relax.

How do we learn to do it ourselves? From directions on the products we use and from how-to-do-it articles in newspapers, magazines, manuals and books.

How many sets of directions can you find near at hand at home? Look in the bathroom vanity, the cleaning closet, the kitchen cupboard. Then read them carefully, and contrast and compare them.

Whatever your field of interest, sooner or later you will have to write directions. Directions are needed for:

- using something, perhaps a hair dryer or a buttonholer
- making something, such as a cookie house
- doing something, folding a napkin, for example

Recipe writing, which may involve any or all of these types of directions, is so specialized it requires a separate chapter; see Chapter 10.

Directions for writing directions

Taking the basic R/C formula—*Visualize, Analyze, Organize, Dramatize* and *Synchronize*, we need to add *Familiarize* at the beginning and, for space reasons, to minimize *Dramatize*.

FAMILIARIZE. Directions often are needed for items that are "hot off the drawing board," so new that only the developers themselves know what to expect.

The development of no-iron wash-and-wear garments is a good example. Laundering specialists working with the manufacturers of clothing made with blends of synthetic and natural fibers had to do extensive experiments with temperatures of both wash water and dryers plus dryer handling before writing the use tags and promotion material. They and their counterparts in the laundry appliance field then had to teach the nation new laundry techniques.

Whether the subject you're preparing directions for is new or not, make sure you thoroughly understand what you must direct others to do. Mix the product, cut with the tool, or tie the scarf until you could do it in your sleep. As you work, look for problems and pitfalls that might frustrate consumers.

VISUALIZE. Will the directions you write have to fit a small hangtag on a dress, a panel on a bottle of floor wax or a couple of inches on a package of frozen fish? Or will they fill a four-page folder accompanying a new vacuum cleaner? Or perhaps they'll be part of a booklet (see Chapter 15 for more details). Fortunately, space for directions (sometimes right down to the size of type and the number of characters per line), is usually specified long before the writer begins to work.

ANALYZE. What does the user need to know to use/do/make that item satisfactorily and safely? What other ideas are nice to know but not absolutely necessary? Would a simple sketch give more information than words?

Break the task into steps. What should be done first, second, third?

Verbalize each step. And I mean that literally: Begin each step with a verb. Can the verb alone carry the message? When General Mills, Inc., discovered that cooks disliked having to sift flour, a new measuring method was adopted. It was summed up in three words: dip, level, pour. It meant: dip the measuring cup into the flour, level off the flour with a knife and pour the flour into the bowl.

DIRECTIONS AND HOW-TOS

If two steps ought to be done simultaneously, don't count on the user having a helper. Figure out how one step can be done a little earlier, or later, without marring the finished product, then insert that helpful word, "meanwhile."

ORGANIZE. Organization in writing may demand imagination. Dream up as many ways to set up your directions as you can. Never assume that the first way you organize the directions—or the way the engineer or the boss does it—is the only way. For example, at one time, garments were constructed first, then decorated. Then a sewing expert took a fresh look at the process and realized it would be much easier to decorate the skirt front *before* the front was sewn to the back.

Organizing your how-to-do-it may mean working out *two* versions: a brief, to-the-point write-up for the user in a hurry, followed by a more complete explanation for the user who will take time to get the whole story. Detergent manufacturers use this approach. The directions are reduced to phrases and printed in large type for the first reader; then, they are repeated in complete statements in smaller type for the second reader.

Once is rarely enough; organizing may have to be done over and over again. Here's a typical sequence:

1. Set up the directions.
2. Try them yourself.
3. Redo the directions.
4. Have a co-worker try them without preparation or prompting while you watch and take notes.
5. Again, rework the directions.

Your goal is to write instructions that are foolproof and failure proof, and couched in common language, with no technical terms.

DRAMATIZE. Because space for how-to-do-it material is usually at a premium, dramatization may have to be confined to such special effects as colored ink for the verb that begins each step, or to typographic devices, such as an arrow or bold dot, preceding each segment.

SYNCHRONIZE. Meeting the printer's deadline is the way you synchronize the publishing of new how-to-do-it materials and directions. More often than any other type of writing—with the possible exception of advertising copy—directions need to be done *yesterday!* Somehow, though, when the boss decides they need to be redone, it's all right to finish them today.

How-to specials for consumers

How-to-select-it articles featured in both consumer magazines and general publications are a more comprehensive version of a how-to-do-it articles. They may deal with items ranging from automatic washers to zippers. And they may involve extensive use/wear tests by a laboratory or other testing organization. Such articles are extremely useful to prospective purchasers who are bombarded by sales pitches and television hype.

These are the elements that a complete *consumer article* (sometimes called a *service article*) will include:

- description of product or item: desirability, uses, background
- quality range
- price range
- availability
- possible pitfalls or problems to consider
- insider's tips
- sources of further in-depth information.

The material about the quality of an item may make up the bulk of the copy (reports on its durability and safety tests) plus discussions of which models might be best for various kinds of consumers. Much of this information may appear in charts and graphs, the construction of which requires great attention to detail. For examples, study any consumer magazine.

Elements of both price and quality are considered and the

DIRECTIONS AND HOW-TOS

best buy—or good, better and best buys on a descending scale—are picked by the researchers.

As electronically programmed features, such as touch controls on microwave ovens, play a larger role in costly consumer goods, how-to-select-it material of this sort will become increasingly complicated, but also increasingly necessary.

10. Recipes and their adaptation. The thinking that goes into a complete recipe

POPULAR cooking instructor Maurice Moore-Betty put it succinctly: "A recipe is a guide, not a gospel."

Keep that saying in mind as you develop skill in recipe writing. Inexperienced cooks may follow your directions to the letter, but those who have been cooking for some time may only follow the general outline, substituting ingredients and seasonings.

Unless you are working for a brand new organization you will be writing recipes in the format your office has been using. It is likely to be the *standard or conventional format* that lists the ingredients, then gives the directions. the directions may be in one or more long paragraphs or numbered in sequence with an action verb beginning each step.

Two other recipe styles or formats, the action and the narrative, are rarely used now, though they were used frequently in the past. The *action style* is chronological, giving the directions first, followed by ellipses, and then the ingredient(s) the direction applies to. This style is especially good for children and other beginners because it is easy to follow.

The *narrative style* is still used by writers who feel it forces the cook to study the directions while seeking the ingredients. A variant of narrative style that is almost telegraphic is often used in advertisements. The same recipe, written in these styles, together with the pressure cooker, microwave and slow-cooker variations of the standard pattern, follows:

RECIPES AND THEIR ADAPTATION

ACTION STYLE

> ### New England Boiled Dinner
> Place in large kettle . . .
> *2 lb. corned beef*
> Sprinkle with seasonings that come with meat. Cover meat with cold water. Heat to boiling, reduce heat and cover tightly. Simmer 3 hours or until tender. Add . . .
> *24-oz. pkg. partially-thawed frozen vegetables for stew*
> Recover kettle and simmer 20 minutes. Add . . .
> *4 to 5 wedges green cabbage*
> Simmer dinner uncovered 15 minutes or until vegetables are tender. Carve beef in thin slices across the grain; serve with horseradish.
> Amount: 4 to 5 servings.

NARRATIVE STYLE. Narrative style may save space when the number of ingredients is short. This style forces the user to read the directions while making the shopping list.

> ### New England Boiled Dinner
> Place 2 lb. corned beef in large kettle; sprinkle with seasonings that come packed with meat. Cover meat with cold water. Heat to boiling, reduce heat and cover tightly. Simmer 3 hours or until tender. Add 24-oz. pkg. partially-thawed frozen vegetables for stew to kettle. Cover; simmer 20 minutes. Add 4 to 5 wedges green cabbage; simmer uncovered 15 minutes or until vegetables are tender. Carve beef in thin slices across the grain; serve to 4 to 5 with horseradish.

ADVERTISING STYLE. A recipe to appear in an advertisement features brand names and brevity.

New England Boiled Dinner
Sprinkle seasonings from 2 lb. REUBEN'S CORNED BEEF on beef in kettle; cover with water. Boil, covered, 3 hours. Partially thaw 24-oz. package STOKELY'S FROZEN VEGETABLES FOR STEW; add to beef and cook 20 minutes more. Add 4 to 5 wedges cabbage and cook another 15 minutes. Slice and serve to 5.

STANDARD OR CONVENTIONAL STYLE. The same basic recipe can be written five different ways depending on which method of cookery is specified. Typically, either the top-stove or the oven method would be given, followed, where applicable and space permits, by one or more appliance modifications. For many recipes there is only one preparation method that is suitable. In fact, only a small number of combination dishes can be prepared successfully all five ways.

The pattern of listing the ingredients first, then the directions, is sometimes called shopping-list style.

New England Boiled Dinner
2 lb. corned beef
1 (24-oz.) pkg. frozen vegetables for stew, partially thawed
4 to 5 wedges green cabbage

Top-stove method: Place beef in large kettle; sprinkle with seasonings that come packed with meat. Cover with cold water. Heat to boiling, reduce heat and cover tightly. Simmer 3 hours or until nearly tender.

Add stew vegetables to kettle. Cover; simmer 20 minutes. Add cabbage; simmer uncovered 15 minutes or until vegetables are tender. Carve corned beef in thin diagonal slices across the grain at a slanting angle. Serves 4 to 5. Excellent with mustard or horseradish sauce. Refrigerate leftover beef for sandwiches.

RECIPES AND THEIR ADAPTATION

Oven method: Follow top-stove method (above), except use Dutch oven with tight-fitting lid. Bake corned beef in 325-degree oven 3½ hours. Add stew vegetables and bake another 30 minutes. Finish dinner on top of stove, cooking cabbage wedges atop other ingredients 15 minutes.

Pressure cooker method: Place corned beef and seasonings from package in pressure cooker; add 1 cup water. Cover securely. Place pressure regulator on vent pipe. Cook 20 minutes at 15 pounds pressure. Turn off heat and let pressure drop slowly. Open cooker and add stew vegetables and cabbage. Cooker should not be filled over ⅔ full. Close cover and cook 5 to 8 minutes, depending on size of vegetables. Turn off heat; reduce pressure immediately.

Microwave method: Pierce beef thoroughly with fork. Place meat in 4-quart casserole or cooking bag set in baking dish. Add ½ cup water. Cover tightly or tie bag loosely with plastic strip cut from end of bag. Microwave at 50 percent power (Medium) 30 to 35 minutes. Meanwhile, cut vegetables into small pieces. Turn meat and add vegetables; cook another 30 to 35 minutes. Let stand 10 minutes to complete cooking before serving.

Slow cooker method: Place stew vegetables in bottom of electric slow cooker. Place beef and seasonings from package on top of vegetables. Add just enough water to barely cover meat. Cover cooker and cook 10 to 12 hours at low setting or 5 to 6 hours at high.

After cooking on low setting, lay cabbage on top of beef. Cover and cook 1 hour longer on low or 30 minutes longer on high. When beef is tender and cabbage is done, slice meat and serve.

The name game

The name of the recipe should be clear and concise, not cute. Readers frequently skim food material, scanning the recipe names. This is easy to do if the titles are set off in large, bold type.

Many recipe names mention prominent ingredients; Oatmeal Cake, for example. Some names denote the food's history. Blueberry Buckle has nothing to do with holding up one's jeans; it's a cobbler dessert from pioneer New England.

Recipes from other nations often need two names: the original and the American translation. This brings up another point: we can learn a thing or two about naming recipes from cooks of other nations. Mexicans, for instance, serve Old Clothes without a blush. It's a beef dish in which the meat is cooked so long that it falls into shreds!

The source of a recipe may lend excitement to its name, particularly if the food can be associated with a glamorous place or a famous person. Montego Bay Banana Bread may seem more intriguing than Grandma's Bread. But don't be tempted to tack on a fancy name when the idea, in fact, is Grandma's.

Decide which aspect of your recipe's many qualities is most striking, then weave that element into the name.

What's in it for me?

The ingredient list of your recipe plays a dual role. First, it is a shopping or check-the-cupboard list. Second, it is a list of specific amounts of particular ingredients. By the time you write up your new recipe, you should have checked, ingredient by ingredient, to see that each is needed, that you know what part each plays.

In developing your ingredient list, be practical. If possible, use the entire unit (can, package, carrot or apple) in the recipe. If the recipe simply cannot accommodate an extra ½ cup of the ingredient, suggest a good and easy use for it in parentheses.

And be specific with ingredients: "1 cup of mushrooms, chopped," is a much smaller amount than "1 cup chopped mushrooms." If you have the information, it's helpful also to insert the actual numbers needed: "1 cup chopped mushrooms (about 8)."

The thoughtful recipe writer includes alternate ingredients

RECIPES AND THEIR ADAPTATION

as often as possible and marks optional ingredients. If the cook happens to be out of green pepper or pimiento, it's helpful to know that chopped celery or carrots might be substituted. And, if children in the household don't like mushrooms, it's nice to know that the dish won't flop if they're left out.

Ingredient cost and diet requirements are other good reasons for suggesting alternates. If peanuts will do in place of costly cashews, say so. When a low-fat cheese, such as Mozzarella, could be substituted for high-fat Cheddar, mention it.

Whether for temperance or convenience, alternates also should be offered for alcoholic ingredients when the flavor of the dish does not depend on them. Fruit juice often is a suitable substitute for sweet wine. Broth frequently may be added instead of dry wine.

The best recipes, many believe, are short—five ingredients or fewer—and easy to remember.

Recipe writers with an eye to the future now include both metric and standard American measurements in their ingredient lists. As cooks buy new equipment, such as liter/quart measuring cups, they can acquaint themselves with metrics, making the expected transition to this worldwide system easier.

Make directions, well—direct

Recipe directions ought to be direct, to the point. Strictly speaking, the writer is telling the cook what to do—in no uncertain terms. First, do this; next, do that. It may seem dictatorial, but it's efficient, from the standpoint of time (the cook's)—and of space (the publication's).

Your recipe's directions are a good test of your ability to visualize and analyze. After visualizing the steps involved, number them (whether you do or do not use a numbered-step format) so that first things come first.

Next, study the steps to see if part of the work can be simplified. For example, if you must beat egg yolks and egg whites separately, write: "Beat egg whites stiff; set aside. With same beater, beat yolks." If you whipped the yolks first, it would be necessary to wash the beaters before whipping the whites because the fat in yolks keeps whites from beating stiff.

Remember two little words: "or until." After giving such vital information as pan size, baking temperature and cooking or baking time, then add "or until" and a description of readiness or doneness. This allows leeway for the variables. Ovens sometimes run hot or cool; so do refrigerators. It helps the cook to know how the gelatin mixture should act when it's ready for whipping, or how the chicken leg reacts when it's tender.

But don't overdo the directions; don't bury your good idea in a heap of unnecessary words.

Another idea: symbols can stand for words. Instead of taking the space to say that the casserole can be completed up to the sixth step, then refrigerated overnight and reheated 15 minutes longer than suggested in step seven, work out a symbol—refrigerator shape, perhaps—that means you can make ahead and refrigerate.

Yield signs

Never end your recipe without telling its yield. Whether feeding 2 or 12, economical cooks like to figure amounts when planning or shopping. Be sure that your kitchen recipe forms have spaces for yields. And, when requesting recipes from outsiders (such as restaurant chefs), provide a form with space for the full recipe, including the number of servings or, for foods such as salad dressings, the amount the recipe makes. When the source does not include yield, make an educated guess based on one-serving "guess-timates": vegetable or salad served as accompaniment—½ cup; casserole or main dish—1 cup; meat, fish or poultry—4 ounces.

Computers assist

Both creators and users of recipes benefit from knowing which nutrients are furnished by a particular food. Now, thanks to computers, we can learn how nutritious a recipe is and include that information.

RECIPES AND THEIR ADAPTATION

Since 1979, the Minneapolis-based Pillsbury Company has made its nutrition data bank available to food departments of newspapers and magazines. By means of a telephone hookup to an especially programmed computer, a recipe can be analyzed for calories, protein, carbohydrates, fat, cholesterol, sodium and potassium. At the same time, the percentage of the National Academy of Science's U.S. Recommended Daily Allowances (RDA) for protein and for major vitamins and minerals is printed out by the computer. To save space, the nutritional analyses usually are published in smaller type than the recipe. Readers who count calories, who must watch their cholesterol or sodium intake or who are concerned about getting enough of a particular vitamin or mineral find these analyses especially helpful.

Something borrowed, something new

Very few recipes are totally original. Most recipes are borrowed from another source, then adapted to different products, simplified or embellished.

Writers preparing recipes for publication can protect themselves from a sticky copyright dispute by doing one of two things. They can credit their source, citing both publication and author; it's not necessary to list page or publisher. Or, they can change the name of the recipe and two ingredients.

Changing two ingredients usually isn't too difficult. Check the seasonings. Perhaps mace or coriander would be just as flavorful as cinnamon or nutmeg? Could you use less sweetener, salt or butter?

Adaptations optional

APPLIANCES. In recent years, many recipes have been adapted to three new kitchen appliances: the microwave oven, the electric slow cooker and the food processor.

Both the zap-it microwave and the set-it-and-forget-it slow cooker help busy cooks save time. And conventional cooking directions are readily adapted to these gadgets. In most cases, adaptation requires adding the correct heat settings and changing the cooking times.

The food processor presents a different dilemma. Because there are many models, it often is wiser to point out that preparation of a particular recipe could be speeded if the cook used a processor, rather than to specify which blade to use and how.

When writing recipe copy that includes adaptations for certain appliances, never hesitate to contact the appliance manufacturers to double-check your changes. The manufacturers' home economists work with the machines daily.

HIGH-ALTITUDE COOKING. Cooks living in mountainous areas require still another type of recipe adaptation. High altitude affects all types of cooking. If your recipe project involves only general comments on high-altitude adjustments, turn to trusted source books. But, if the project requires detailed directions for high-altitude cooks, you might try consulting Colorado home economists who have studied this type of cookery.

MAKE IT DIET RIGHT. You also may want to consider adapting your recipes for particular medical diets. Interest in calorie-reduced diets has been high for some time and interest in salt-, sugar- and fat-reduced diets seems to be increasing steadily. Other cooks may want fiber-rich recipes. When adapting recipes for medical purposes, it's best to consult a dietitian, so you can rest assured that you have made the right substitutions.

For further reading

Fisher, M. F. K. "The Anatomy of a Recipe." In *With Bold Knife and Fork*. New York: Paragon Books, 1968.

Handbook of Food Preparation. Current edition. Washington, D.C.: American Home Economics Association.

Methven, Barbara. *Recipe Conversion for Microwave*. Minnetonka, Minn.: Publication Arts, 1979.

11. Meal plans and menus. The supporting role that menus play

JUST as pearls go with basic black, and crystal goes with damask, so sweet potatoes go with ham and cider with doughnuts.

Yes, knowing what goes with what is an important aspect of the fields of food, fashion and furnishings. Decorators suggest tables to go with sofas and chairs; fashion coordinators pick scarves and jewelry, even hosiery, to complement suits and dresses.

But the meal planner must come up with the most combinations. Call them "go-with," "bake-alongs" or side dishes. Home cooks like to know what goes with what and so do restaurant cooks and customers. Menus that evoke appetizing images are vital as well to the success of the hospital dietary department, the school cafeteria and the white-tablecloth restaurant. And they're important in nutrition education and in the food business, too.

Balancing act

Writing menus and meal plans demands a wide range of skills.

First, know the principles of meal planning: how to balance a meal's textures, temperatures, colors and calories. If you have not studied meal planning, look for a menu cookbook or a recent meal-planning manual. Begin a file or scrapbook of intriguing menus clipped from publications.

Second, develop a vocabulary of food words. Beg, borrow or buy a stack of national food magazines, then scan the pages, making a list of the standout food phrases. The writers of this copy are highly paid, competitive and talented. Collect menus, too, whenever you eat in a restaurant. What some proprietors put into print will amuse—or amaze—you.

Third, learn all you can about the foods you will use in menus. This may mean preparing some of the foods yourself so you can rely on personal experience to label foods "easy-to-make" or "child-pleasing." Or, you may need to hang around the test kitchen to get samples of new products or recipes that you'll be suggesting accompaniments for.

What makes it delicious?

Many, many foods are delicious. But the conscientious writer strives for words that describe the dish's eating quality more precisely and, at the same time, help sell the menu item. Is it buttery, chewy, colorful, concentrated, cool, creamy, crisp, crunchy; crusty, dainty, delicate, delightful, elegant, filling, fragrant, fresh, frosty, fruity, hearty, intriguing, juicy, mellow, nutty, piquant, pungent, refreshing, rich, robust, satisfying, savory, spicy, sumptuous, superb, sweet, tangy, tart, tender, wholesome or zesty?

Beware of such words as "special" and "exciting" which have been greatly overused. Beware, too, of piling on too many adjectives.

Menus for publications

Including menu plans in publications is an ideal way to increase the usefulness of the materials your organization is preparing. A recipe booklet can mention an accompaniment for each food, making the booklet doubly helpful. A new-product release can describe how the item can be served for different occasions or during different seasons. A newsletter for cooks or for owners of a particular appliance can showcase ideas for combining featured foods for snacks, parties, even gifts.

Menus used in a cookbook or feature article are particularly helpful to two types of readers: those who have absolutely no idea of what to serve with what, and those who are in a

MEAL PLANS AND MENUS

rut, always serving the same foods together. The first group will clutch any helpful straw; the second needs a bit of excitement to entice them away from their comfortable rut. Try to include both easy and elegant menu combinations in the materials you write.

Adding menus is an excellent way to take advantage of space on a cookbook page that is too small to accommodate another recipe or variation. If you have enough room, include a entire menu featuring a recipe on that page, or better yet a menu made up of a recipe from that page plus recipes found elsewhere in the book. Another time you may have only enough space to suggest what to serve with the featured recipe, perhaps just the right salad to round out a casserole supper.

Whatever the format, mentioning a "go-with" food makes any recipe in print "taste better."

Menus for restaurants

Food service is one of the fastest growing and most dynamic segments of the food business. Each year, statistics show, more and more meals are eaten away from home. As a food professional, you have a dual purpose when called upon to help write a restaurant menu. The first is doing your share to build a well-run, profitable eating place; the second is keeping in mind the needs and desires of its customers.

Let's say that the owner of the restaurant chain has devised a new sandwich that he wants to promote heavily. It's your job to suggest accompaniments. Potato chips and a dill pickle have been the time-worn accompaniments in the past, but these high-sodium foods are being shunned by the health-conscious customer. "How about a fruit garnish instead?" you ask yourself. It *would* be better nutritionally. So, you consider the objections your boss might have: time and money. When you suggest the fruit garnish, you show him a sketch of the finished plate and follow up with all the dollars-and-cents details of cutting, storing and arranging the garnishes. Management likes the idea, not only because it's a fresh suggestion but because you have anticipated and worked out all the kinks.

Perhaps you or a group of your friends will launch your

own restaurant someday. The menu is the key. A soup bar has different considerations, from procurement to presentation, than does a natural foods eatery. You cannot offer six kinds of homemade breads when the baker has time and oven space for only three.

What's more, menu offerings must be crystal clear. The menu reader (who may be both hungry and tired) shouldn't have to wonder which entrees include the salad bar and which do not. It may take clever use of parentheses and/or asterisks, but by reading your carefully-written menu the customer should be able to understand restaurant policies on smoking, tipping and substitutions.

Remember, too, that today's restaurant customers are pretty sophisticated. Don't label an item Eggs Benedict unless it is indeed the New Orleans specialty replete with Hollandaise sauce. If your kitchen is topping the poached eggs with cheese sauce, rename the specialty (for the cook, perhaps) rather than setting the diner up for disappointment.

Imagination *and* accuracy are musts in writing food descriptions for restaurant menus. Your adjectives must attract interest without conveying false impressions. A truth-in-menus movement seems to be under way. Government agencies and consumer activists are calling "fraud" when menus promise one level of quality but a lower one is served (such as canned or frozen vegetables instead of fresh).

Menus for other food service institutions

While most restaurant diners choose where they'll eat, those who must eat in company or school cafeterias and in hospitals and other such institutions often can't or don't. For them, variety is the spice. And you, as the menu writer, can work hand in glove with the food service manager to increase variety whenever possible.

Here are three examples of how a thoughtful writer contributed to an orgaization's food service:

• The assistant manager of a company lunchroom was responsible for writing the daily lunch menu, which, in turn, was posted on bulletin boards throughout the plant. After studying the menus that had been used the preceding months, the assistant remembered a trade magazine that had listed special weeks and months promoting particular foods—for an entire year. For example, June is called Dairy Month by the American Dairy Association. So, the lunch menu featured ice cream one week, cheese another, etc., while posters from the dairy association brightened the lunchroom.

- The social director of a convalescent home doubled as the menu writer for the dining room. Because the social director knew all the residents well, she could pass on to the cook a favorite recipe of a particular resident. When the cook made the cake or salad (or whatever) for the home's residents, the first serving was presented with a bit of fanfare to the one who provided the recipe, and the menu featured Jane Jones's Honey Cake.
- The cook-manager of a junior high school wanted to give the students' meals a new twist. So, working with the principal, she formed a student advisory council made up of boys and girls selected by their classmates. The council decided to link the food with subjects the youngsters were studying. When Russia was the subject in geography class, Hamburger Stroganoff (creamed hamburger with a bit of sour cream) was served. And when the Spanish club was putting on a play, the cafeteria dished up a special flan (baked custard).

For further reading

Burros, Marian. *Keep It Simple, 30-Minute Meals from Scratch*. New York: William Morrow, 1981.

Dannenbaum, Julie. *Fast and Fresh: Delicious Meals to Make in an Hour (or Less)*. New York: Harper and Row, 1981.

Kotschevar, Lendal H. *Management by Menu*. Chicago: National Institute of the Foodservice Industry, 1975.

Restaurants & Institutions, semi-monthly, published in Boston.

12. Features and columns. The six types and how to do them

FOR BALANCED READING, editors offer both news stories and feature articles. In writing a feature article, you are spotlighting an item, an event, a personality or a project.

A newspaper or magazine feature section usually is divided between articles and columns. Feature articles are impersonal—though they may not be totally objective—while columns are highly personal. The feature writer uses third person—*he, she, it,* or *they*; the columnist uses first person—*I.* Columns often run a single column wide, hence the name.

As a beginner, you will write features, both of a general nature and relating to your specialty. After gaining maturity and attracting a loyal audience, you may be asked to write a column. Occasionally, on smaller publications, the writer covering life-style, food, fashion, housing or interior design will prepare one or two longer feature articles, then add a column of short personal observations, reactions and helpful hints.

For a comparison of columns and feature articles, scan a current women's magazine. The format calls usually for placing columns first; feature articles and illustrations in the center of the magazine.

Many columns, you will notice, are limited to a single subject: microwave cookery, family finance, jobs for women, health. And the columnists are established experts in their fields; many have written books on their subjects. Other columnists represent an unusual viewpoint, that of a house husband, for example.

Most feature articles fall into one of these six categories:

personality, problem solving, trend/life-style, backgrounding, how others live and seasonal.

I'd like to talk with you about . . .

Whether the subject is *haute couture* or *haute cuisine*, interviews with experts yield insights.

Food writers frequently interview outstanding local cooks, chefs, caterers, product development experts, cookbook authors, nutrition researchers, dietitians and food scientists. Fashion and furnishings writers talk with designers, manufacturers and store buyers. Writers who specialize in the family seek interviews with psychologists, leaders of family organizations and outstanding parents and educators. Consumer writers, in turn, interview product designers, consumer representatives and government regulatory officials.

It may take some enterprise to find knowledgeable people to interview. But leaders in a variety of community organizations can direct you to people who would be willing to share their expertise. These organizations would include state and county home extension services, office of the mayor or city council, restaurant and resort owners' association, craft clubs (such as the Weavers' Guild), the Chamber of Commerce, the Council of Churches, the local chapter of The Fashion Group, the senior citizens' council, the Senior Citizens Center, the Mental Health Center, the local labor federation or the international center at a nearby college or university.

Here are some time-proven guidelines for successful interviewing:

- *Set up your interview by telephone.* Explain just what informational territory you would like to cover and how long you expect the interview to take.
- *Do your homework.* If the person is a visiting expert whose interviews are being set up by a public relations office, you can obtain a biography from that office. If the subject is nationally known, learn about her/him in *Who's Who* or *Current Biography*, from clipping files in the public library or from files at your local newspaper. This saves time because details of birth, schooling and family need not be

repeated during the interview.

If the person is promoting a new book, get it and skim-read it.

If the subject of the assignment is not familiar territory for you, take time to read up on it. For example, few young food writers have done much canning and freezing, yet they may need to interview an expert about streamlining those food preservation techniques.

If you want the interviewee to comment on current events or controversial findings, review the details so you can understand her/his remarks.

- *Prepare questions in advance.* Be ready, too, to abandon your line of queries if your interviewee launches into an attractive subject area.
- *Arrive on time for the interview.*
- *Give sincere compliments—but not flattery—to elicit good feelings.* If you have admired your interviewee's work, read her/his book or enjoyed a show featuring her/his fashions, say so. This helps break the ice.
- *Ensure the accuracy of the interview.* You might decide to tape record it. However, while recording devices free the interviewer from note-taking and help to ensure accuracy, the intrusion of the microphone may lessen the spontaneity of your conversation. Also, the recorder may malfunction, and then, too, transcribing the tape is time-consuming.

Many would-be writers learn shorthand or speed writing so they can take notes, yet keep eye contact with and respond to their subject. Some writers make notes only of key quotes but have trained themselves to concentrate hard during an interview. Immediately afterward, they write down what was said, together with observations of the subject's demeanor, dress and surroundings.

Get the exact spelling of the interviewee's name, the precise wording of her/his title and, if applicable, the complete titles of books and articles. Be sure to do this before the give-and-take of questioning.

- *Be professional in your questioning.* Ask questions in a business-like manner, but don't fire them too rapidly.

Let the conversation develop naturally. Mark X's in your

notes where you want to expand or clarify a statement, then remember to reopen that topic as the interview tapers off.

Don't permit a subject to recite the same carefully calculated statements repeated to every other reporter across the nation. Beware the celebrities who can put their tongues on automatic pilot and dictate an article about themselves without flicking an eyelash. Interject questions and comments until you steer the interview into unexplored territory.

Guide a person who strays from the topic back to the material you want to cover. Some interviewees love to reminisce, but, in doing so, they digress.

Rephrase and repeat a question that has been circumvented or given short shrift by the interviewee.

Wind up the interview quickly after you have covered the agreed-upon material. People who are important enough to be interviewed usually keep tight schedules. Thank your interviewee for her/his time and mention (if you know) when the article will appear; offer to send clippings.

- *Write your impressions immediately.* Interviewing a celebrity or expert in a field of high interest for your audience can be exhilarating. Before that excitement wears off, write down some impressions from your meeting: how the person gestured, what s/he was wearing, how the surroundings reflected the person's life and interests.

If you established good rapport and the interviewee was fluent and knowledgeable, you may return to your desk with enough material for three stories. But you'll write only one. If you select details and summarize information with care and thought, your reader can enjoy meeting the person as much as you did.

Creative problem solving

Articles that focus on problems involving money, quality of life or both seem to make the most interesting problem-centered features. Facts form the backbone.

Here are some examples in telegraphic form:

Problem: family has moved into large ultramodern home

and needs furniture; wife inherits a lot of antique furniture. Solution: wife upholsters antiques in sharp modern fabrics.

Problem: professional woman with high-visibility job, pregnant for the first time, needs maternity wardrobe. Solution: selection of quality dresses in straight-cut, loose designs, which can be worn belted after baby arrives.

Problem: no money to pay baby-sitter. Solution: trade child care with a friend, or barter work you enjoy for child care.

Problem: son must leave for work before parents return for evening meal. Solution: family reorganizes budget and meal planning to take advantage of the microwave oven and electric slow cooker.

You, the writer, must be on the lookout for anecdotes about problems others have faced and solved. As you converse with colleagues and friends, your news gathering sense must respond: there's a story!

Attend to trends/life-styles

As our life-styles evolve, we move with the trends. This continuing process presents many possibilities for astute and observant writers. For example, consider the gradual transfer of responsibility for housework from the homemaker alone to the family as a cooperating unit. Introducing your readers to people whose lives are changing, and describing their reactions to those changes, makes fascinating reading.

Color in the background

All feature articles include *background*, that is, information vital to understanding a problem or situation. But, occasionally, an entire article is devoted to the background of a subject. Perhaps your main article is a quality-and-cost comparison of leading brands of ice cream. Your *companion article,* or *sidebar,* might deal with the history of ice cream or describe a visit to an ice cream factory.

Or, let's say your top feature is a fashion show by a leading designer. A natural background piece would be a biography of the designer, tracing early influences on her/his creative muse.

Walk a day in another's shoes

In a vast nation, a far-flung world, we cannot meet personally all the people we would enjoy knowing. Foreign exchange visitors, students or adults, provide good sources for how-others-live articles. Whatever your interest—housing, food, clothing or family life—ask the visitor to discuss that aspect of life in her/his country. Another possible source is your city's international center or organization that helps new residents adapt to this country.

Readers also are intrigued by the daily lives of those with unfamiliar occupations or living situations. During the 1960s, when young people were eschewing the establishment, forming communes and going back to nature, many articles detailed the shared work and child-rearing in communes.

Many readers are somewhat familiar with grain and livestock farming, but what about bee farming or mushroom farming or managing the most glamorous of all farms, the vineyard?

Or, consider the historical how-others-live article. It can be based on an interview with a keen-minded old-timer who remembers when main street was a cow path or who can describe the local dignitary for whom the junior high is named.

'Tis the season to be . . .

The seasonal feature story is repeated again and again, but in different guises, of course. Each fall, fashion writers describe back-to-school clothes. Each November, food writers discuss the sumptuous Thanksgiving feast. And each January, there's usually the psychologist being interviewed about the post-Christmas letdown. At income-tax time, writers suggest ways to economize. The same is true for Christmas, Easter, springtime graduations and bridal showers. You must know the sequence well.

Many communities have an annual festival that provides

opportunities for seasonal features: the selection of the queen, the parade, the dance contest, etc.

Getting a foot in the feature-writing door

Nowadays, many short-staffed publications encourage the efforts of new writers, although they may not pay much. If you want some experience, pack up samples of your writing, put together a list of story ideas and make a date with the editor.

Here are some feature possibilities that are close at hand, yours for the researching.

- How a handicapped student manages to keep up with fast-paced college life; these people usually reveal courage and high ideals.
- An interview with a celebrity or a civic leader who is brought to campus for a seminar or speech; get her/his predictions for future developments in the field.
- A behind-the-scenes look at something everyone takes for granted, from the air-filtration system of the auditorium to the costume storage for the theater.

If you are a homemaker and aspire to be a part-time writer, you can write articles based on at-home experiences. Many of these fall into the problem-solved category. These suggestions may ring a bell with you—and with an editor.

- Enforced ingenuity: what you did when you arrived in your new home but your new range did not arrive; how you cooked and baked with an electric frypan and a popcorn popper.
- Recycled fashions: ideas—with sketches—for adapting and adjusting outgrown or worn children's clothing for another use or another season.
- Easy amusements: how to keep toddlers happy in a doctor's office or on a car trip. Collect ideas from other young mothers.

Columns don't come easily

If you yearn to be a columnist, you should isolate a subject area early in your career and become an expert in it. It may take years to become knowledgeable enough about a topic to advise others on it. For example, a teacher became a wine columnist at the seemingly early age of 29. But he had taken a serious interest in wine at age 21, had participated in tastings two or three times a week, and had traveled to wine-growing areas in the United States and abroad before approaching his local daily newspaper about writing a column.

Careful preparation is also a must for the writer of a restaurant review column. This writer, like the theater reviewer, has the ability—the power—to make or break a business. Therefore the restaurant reviewer should be diligent in research and scrupulous in fairness.

Such a columnist is a specialist, an expert, an insider. Other types of columnists are name-droppers, participant-observers and those who have the view from the top. The name-dropper columnist is a civic-minded cousin of the gossip columnist. Each column carries as many names as possible of authors, entertainers, professors, vice presidents and the like.

The participant/observer columnist often writes in conversational, almost confidential terms, discussing the doings of the family and the neighborhood, but always with an attitude of universality.

The view-from-the-top columnist is the president or chief executive officer who decides to pass along her/his observations and wisdom in an industry or company publication. Because these people often are extremely busy, they may ask someone on the staff—it could be you—to ghostwrite all or part of their columns.

For further reading

Brady, John. *The Craft of Interviewing.* New York: Vintage Books, 1977.

Rivers, William L. *Finding Facts, Interviewing, Observing, Using Reference Sources.* Englewood Cliffs, N.J.: Prentice Hall, Inc., 1975.

Rivers, William L. and Shelly Smolkin. *Free-Lancer and Staff Writer, Newspaper Features and Magazine Articles,* 3d ed. Belmont, Calif.: Wadsworth Publishing Company, 1981.

13. News releases and public relations.
The challenging tasks of the publicist

A NEWS RELEASE is a press release is a publicity release. Whichever label you give it, a *news release* is a brief piece of writing designed to bring a product, a method, an idea, a person or a group to the public's attention.

Many releases are published with little or no rewriting; others may be combined with additional information in a staff-written article.

The publicist works in the complex area of communication known as public relations. As a publicist, you form an important bridge between the organization for which you work and the media. To do an effective job, you must be thoroughly familiar with your company (or association or institution) and the newspapers, television stations and magazines that receive your releases.

The news release is the tool of the publicist, a writer who specializes in short articles submitted to news media on behalf of an organization. Even those who have never seen a news release have undoubtedly read many of them.

Most news releases about food, families, fashion and furnishings are prepared for daily newspapers. A generation ago, newspapers were rigidly departmentalized. The women's page or home section carried recipes and other articles of interest to wives and homemakers. Now, however, newspaper sections are much broader based and may bear such wide-open titles as "Life-style," "Leisure" or simply "Food."

The editor of such a section has one purpose: to pack the columns with useful ideas in words and pictures, ideas that are fresh—if not absolutely new.

Most editors have space for only a few of the articles and pictures their secretaries pile on their desks each day. So they quickly glance at the titles, topics and sources of the releases, tossing those that ring a bell into a basket for further study and consigning the rest to the files, round or otherwise.

You must realize that your release will be competing with dozens of others for the editors' eyes before it even has a chance of reaching the readers. Thus you need to know all you can about what editors usually publish—which, of course, reflects their judgment of what their readers like to read. (Yes, the word is audience. See Chapter 2.)

Reason-able releases

The writing, printing, photography and postage involved in the design and mailing of news and publicity releases are so costly that the idea must merit use by at least some of the media.

More often than not, a release is part of a larger promotion by a company. These sorts of situations prompt publicity efforts:

- a new product or line of products is being introduced
- a large supply of a product is available
- a holiday or season for which the product is appropriate is coming up
- a new use has been devised for an existing product
- a life-style change is occurring in which the product can be helpful.

Some of the recent life-style changes that have occasioned much written material include: the increased number of singles, the return of many mothers to the work force, the demands of inflation, the decrease in family size and the energy crunch.

Let's study a hypothetical series of releases and the promotion of which they were a part. The promotion is a joint effort by two organizations: a spice company and a housewares maker. The two decide to *tie in* (combine their efforts) on a vegetables promotion.

NEWS RELEASES AND PUBLIC RELATIONS

The creative people working for the companies have noted that meats are getting more expensive and that polls show consumers are serving more meatless meals. The housewares manufacturer recently introduced an easy-to-use vegetable steamer and wants to follow up introductory publicity with a second campaign featuring steamed vegetables enhanced with herbs and spices.

The staffs of the two companies and of their advertising agencies search their files for suitable recipes, then begin developing and adapting ideas. Finally, eighteen steamed-vegetable recipes that fit the theme are presented to the writers and account executives. From this array, one vegetable platter is picked as the subject for the advertising campaign and six others are selected for publicity releases. Twelve of the recipes are to be included in a free booklet to be offered in the advertisement.

The ad and the series of releases are both necessary. Media experts know that some readers will read one, some the other, and a few will see both. The publicity recipe collection includes herb-seasoned vegetables especially for six particular groups: single cooks, home gardeners, working couples, patio party givers, calorie-counters and gourmet cooks. Thus, each release has a built-in angle.

The big four

All news releases should carry four key elements preceding the copy and illustration: source, release specification, statement of exclusivity and title.

SOURCE. You and your organization are the source, of course. The full name of the company (including the address and telephone number) usually is shown on the letterhead or form the release is printed on. Your name and telephone number should be shown in the upper left hand corner of the page. Often your supervisor's name or the name of another writer familiar with the product is included with yours, just in case you are busy when an editor calls to verify information or ask for help. If a public relations agency handles information for several clients, the name of the client for whom the release is being sent is listed as part of the source.

RELEASE SPECIFICATION. The release specification shows when the material can be released by the press. This specification frequently appears in the right-hand corner of the release. The phrase "For Immediate Release" is used on most releases involving homemaking products. However, when the information is highly competitive or when a product will not be distributed until a certain date, the release should say: "Hold for Release," then specify a date. Editors and writers usually are careful to abide by release dates.

It also is wise to include the day and the year a release is issued. Many food and life-style editors keep releases, particularly those with recipes, on file from year to year. Editors like to know which releases are most current.

STATEMENT OF EXCLUSIVITY. You may want every editor in the country to see your release, but editors take a different view. They do not want to open another newspaper in their city and discover that the timely party plan they were planning to feature next week has been used by a competitor. To avoid this, publicists prepare a variety of releases for the same product and furnish different articles to editors in the same readership area. They then mark the release "Exclusive to you in your area."

For a frequently used product, nonexclusive releases are acceptable. These might be marked by the sender, "Good, but not exclusive."

TITLE. The title of your release should be centered about one-third of the way down the page.

Because the title is the key to the release, it should be chosen carefully. In some cases, a catchy title, much like those used in magazines, is in order. For example, "Surprise Packages" is a perfect title for a release describing foil-wrapped, ready-to-bake individual servings.

Another time a news headline is more appropriate. "New Model Joins Food Processor Family," for example. It topped a piece about a major appliance manufacturer introducing a new processor.

Sometimes a label title, such as "Freeze-Ahead Potato Dishes" or "Easy-Sew Spring Accessories," is satisfactory.

NEWS RELEASES AND PUBLIC RELATIONS

Another time a direct question may work. Your title can ask the editor: "What's New in Children's Play Equipment?"

Occasionally, a question-and-answer approach can be taken. This one, in fact, is a mini-lead: "Italian Pizza? The Spanish Created Their Own...with Onions."

Will your title tease or will it tell? It's up to you. While clever wording and literary devices often are used in release titles, good judgment must prevail. Don't let a linguistic device, such as alliteration, cause you to stumble. A writer for a pickle company overdid it in this title: "Pickles Provide Palatable Pleasure."

Good, bad or indifferent, don't expect your title to appear in print. Copy editors write the headlines that accompany such articles in magazines and newspapers.

Attention: editor

The release title attracts attention, but the idea you present must hold it. The information in the release and the illustration that accompanies it must fulfill the promise of the title.

Release copy should be short and to the point. The two-thirds of the page below the title usually is the right amount of space. Let the reader know why the idea (or product) is good, then move right into the details of putting it into practice. Releases have much in common with the feature articles discussed in Chapter 12.

When you spend a lot of time working with a product or appliance, you become enthusiastic about its possibilities. When that happens, it is easy to let superlatives slip into your copy. In fact, some release writers tout their subjects as "best," "finest," "newest." Sometimes the editor will like your angle well enough to edit out the superlatives, but other times such lavish language can cause your release to be rejected. It's best to avoid *puffery,* the practice of puffing up your story with boasting language.

Publicity writers hate to admit it, but the illustration—not the writing—catches the editors' eyes. A good, sharp photograph of an easily recognized subject often is more valuable to life-style editors than the freshest of ideas. Photography is time consuming and costly, and many feature writers are more comfortable with words and phrases than with setting up products and/or models and props. Therefore, invest as much thought and planning in your photograph (details in Chapter 10) as in your writing.

Kit 'n' kaboodle

News releases are the key ingredient of a *press kit*, a portfolio containing an introductory letter, several pages of copy, illustrations and, often, a fact sheet and a question-and-answer sheet on the product. The history of the product, a booklet featuring it, and fillers mentioning it sometimes are included. Companies that want to make a splash with their kits aim for pizzaz. A kit featuring aluminum foil for freezing might arrive in a portfolio decorated with snowflakes; releases touting a new strawberry-flavored dessert could be printed on stationery overgrown with berry plants.

A *fact sheet* is a chart listing: product description, generic name, size specifications, suggested retail price, yield, distribution area and shelf life and storage.

The *question-and-answer sheet* could contain five or six general queries, the sort of thing a typical reader would wonder about when considering a new item.

The big event

When a major national venture is being launched—an entire new product line, for example—the press kit is only part of an all-out press event. A press conference, a talk by a well-known speaker, a cocktail party, a dinner or an outing may be the focus of the event, with representatives of the print media, television and radio as guests.

A dress manufacturer, fabric mill or sewing store chain might stage a fashion show with a buffet afterward for the press. A wallpaper manufacturer or interior design company might present a slide show of soon-to-be-announced products narrated by the designer. A cocktail party later would offer an opportunity for reporters to meet and question the designer and other company officials. This one-to-one contact allows writers to individualize their articles with quotes and observations from those they interview.

For further reading

Leiding, Oscar. *Layman's Guide to Successful Publicity.* Bala Cynwyd, Pa.: Ayer Press, 1979.

Tedone, David. *Practical Publicity, How to Boost Any Cause.* Boston: Harvard Common Press, 1983.

14. Magazine articles. The work that brings beginner's luck

> Wanted: Writers with ability, enterprise and persistence. Send query to the magazine nearest your special interest.

YOU can answer this ad now or later, as a student, as a full-time professional or as a homemaker. Free-lance magazine writing is an ideal way to expand your job horizons, make money and remain active in your field.

As any regular magazine reader knows, magazines have changed radically and continue to be a dynamic medium. While some widely known national magazines have folded and others have undergone major changes, many new magazines have appeared. In this age of specialization, there now is a magazine for most hobbyists, the owners of all sorts of equipment, the members of major organizations and the customers of many big businesses.

These specialty magazines often buy articles from free-lancers rather than hire a passel of staff writers. Although the use of free-lancers usually is dictated by economics, it offers advantages. In some cases, the writer already is an expert on a topic that would require weeks of research by a staff writer. In other cases, the free-lancer can devote total attention to a story without the interruptions of office duties.

As a would-be free-lancer, you need to understand how magazine articles differ from the features and columns discussed in Chapter 12. In general, additional depth and quality are expected of a magazine piece because more space is devoted to fewer topics.

Halfway up and halfway down

Magazines, you see, fall midway between newspapers and books in the print media hierarchy. While a newspaper is read, then tossed or used to line the bird cage, a magazine frequently is kept for two months or longer and may be shared with another household. Magazines are perceived as more reliable than newspapers, which can correct or expand an item the next day. The editor must see that the magazine merits readers' reliance by checking the accuracy of each article thoroughly.

Many of the topics discussed in this book apply to magazine writing. Coming up with a provocative idea (Chapter 4) is one; putting your best writing forward (Chapter 5) is another. Being keenly aware of your audience (Chapter 2) is particularly important because readers differ from magazine to magazine. Here's a hint from a midwestern magazine editor, "To know a magazine's true audience, study its advertisements."

In the marketplace

Marketing and querying apply specifically to magazine writing. *Marketing* your feature article means seeking out the editor of the magazine most likely to buy your article. If you are a novice, "buying" actually means publishing, giving you a display window for your talents, a byline for your scrapbook. After you have gained experience, buying a piece of writing involves a check, either on acceptance or when it is published.

There are two possible routes to making a sale in the magazine marketplace. One begins with the topic and leads to the magazine; the other begins with the magazine and leads to the topic. Let's trace both with typical family features.

BEGINNING WITH THE TOPIC. A suburban homemaker-writer spotted some beautiful ski caps while outfitting her children for the winter. She learned that the caps were produced by a talented handicapped woman using a knitting machine. The writer interviewed the woman and noted photo possibilities.

After working out a catchy lead, she queried a magazine for working women. When the magazine editors answered that they had all the articles they could use, she queried a magazine devoted to handwork created in the home and sold the article.

BEGINNING WITH THE MAGAZINE. A college student, who happened to be living in an apartment for the first time, analyzed a magazine edited especially for young apartment dwellers as part of an article-writing course. Eager to make a sale and impress her instructor, she quizzed other young people in her building about situations and activities. Her research led to an information-packed article on selecting secondhand furniture, and, later, to both a modest check and an A in the course.

Once you zero in on a topic that has the depth and breadth to make a good magazine article, study a marketing guide such as *Writer's Market,* then head for a library to study recent issues of the magazines you hope to sell to.

List the magazines that might want such a piece, beginning with *top dollar* (highest rate per word or per article). After the piece is finished, mail it to the magazine at the top of the list. If it is returned, send it along to the next editor on your list, being careful to remove the letter or rejection slip, of course.

So your zeal will not flag, follow the recommendation of experienced writers and line up a family member or close friend to handle the routing of returned manuscripts.

A query is a question

The key to the sale of your idea and, ultimately, your article, is your *query*, the letter designed to persuade the editor to accept your idea.

The query system has long been used because it saves writers and editors time. There's no reason to research and write an article that no one wants to print, or, worse yet, that someone already has printed.

A successful query usually contains four specific elements:

- a lead that states the topic while "hooking" the editor's attention
- a reference to your personal concern with the topic
- further material to excite the editor's interest
- a statement of your willingness to write on speculation

All this must be covered in a single-page letter.

If your lead doesn't grab the editor (whose interest is motivated by the need for material), how can it claim the attention of the future reader, whose interest may be only casual?

Put all the thought, effort and ingenuity you have into your lead and the rest of the query just to get that all-important green light. In fact, longtime editors admit that an appealing, thought-provoking lead will prompt them to consider a topic they were only marginally interested in.

Mentioning your own interest in the topic gives the editor a glimpse into your personality. If you reveal that you collect vintage millinery, for example, it shows the editor you have a strong personal interest in your suggested article about the revival of designer hats.

Just how much more material you will be able to include in your query letter often depends on how much research you have done. If the research is complete, say, for a term paper or for your daily job, by all means outline the intended article. Let the editor see how you would flesh out your topic. Or, perhaps the best you can offer is the same glimpse of the subject that piqued *your* interest.

As a beginner, you will, of course, be happy to write on speculation—that is, completing the article even though the editor is not committed to buy it. You're gambling that your story will be good enough for publication. Novices may work on speculation for a long time before their work gains enough recognition to prompt assignments from editors. In writing, as in so many other fields, beginner's luck in magazine writing depends on the beginner's work.

You've invested a lot of time and energy in your query, so don't fail to observe query protocol. These purely mechanical requirements are musts:

- enclose a stamped, self-addressed No. 9 envelope along with your query in a No. 10 envelope
- use a pica typewriter—not elite
- use quality bond paper, not the easy-erase kind that smudges
- write the current editor at the correct address—telephone, if necessary, to ascertain this
- have the letter spaced and typed *perfectly*—sloppy is as sloppy does.

Seek and ye shall find

When the editor lives in another city, you must submit yourself to the printed page, then await the answer of the editor or her/his assistant. This process often takes weeks, sometimes months. And what does the hopeful writer do in the meantime? Research other magazines, dream up other topics, write queries to other editors.

For free-lancers, the adage has been amended: all things come to her—or him—who *hustles* while s/he waits.

Much can be said for face-to-face contact with an editor who might, just might, be interested in your articles. Seek out editors in your community and contact them for an interview regarding free-lance work. If you have a network (friends who support you personally and professionally), seek suggestions of editors they may know.

Here are some other magazine possibilities: a company publication where you or a member of your family works, the entertainment magazine published in your city for convention visitors and tourists, the public relations magazine of a business you patronize or might patronize—from your life insurance company to your community hospital.

Want more? Look in the yellow pages of your telephone directory under Publishers—Periodical. One fledgling free-lancer, who had little background in the wall-covering industry, got acquainted with the topic (from a friend who was an interior designer) when she learned that a locally published wall-coverings trade journal needed help. She earned confidence—and cash.

For further reading

Boeschen, John. *How to Make Money Freelancing, a Guide to Writing and Selling Nonfiction Articles.* Richmond, Calif.: Wordworks, 1979.

Provost, Gary. *The Freelance Writer's Handbook.* New York: Mentor Books, New American Library, 1982.

Writer's Market, annual, published by *Writer's Digest* magazine.

15. Booklets. The whats and whys of folders and booklets

BOOKLETS: how we love to pick them up at workshops and conventions and at stores. Colorfully printed, brightly written, they're short enough to read at one sitting, and small enough to slip into a file for future reference.

Preparing booklets, bulletins, pamphlets and folders presents the writer with a challenge and an opportunity because the subjects can be so varied.

Booklets must be small, as the name (little book) implies. But how small? Purse size? Number 10 envelope size? Chances are, the booklets you have saved range from 3-x-5 inches to 6-x-8 inches. The number of pages ranges from eight to thirty-two (the number of pages, by the way, must always be a multiple of four).

Because a booklet *is* small, its topic must be limited in scope. Home-sewn decorative accessories may be the rage, but stick to one for your booklet, say, pillow-making.

Long before a topic can be selected, the booklet's purpose must be decided upon. Organizations publish booklets for one of three reasons: as a service to customers, as a promotion for a product or as a gesture of goodwill. For example, when the energy pinch began to be felt, utility companies speedily put together pamphlets packed with suggestions for conserving fuel, and mailed them to customers along with monthly statements.

Once the purpose is plainly stated, you (or your production committee) can move to the format. So many factors influence format that they need to be considered one by one:

- shape (often determined by the other factors)
- distribution method
- color or black-and-white

BOOKLETS

- photos or sketches
- free or nominal charge
- single or series

Shipshape

If your booklet is to be mailed, it should fit inside a standard mailing envelope, preferably the 6- or 10-inch length. But if the booklet is to be distributed at a convention or used as a table favor at a banquet, it can be tall and slim or short and squat.

Booklets sometimes stem from other printed material and that may influence the shape. For example, the labels of a vegetable canner featured a picture-illustrated recipe for each vegetable. When cooks began asking the company for copies of earlier recipes, a booklet was designed teaming two can label recipes, side by side on a page. This trick saved a great deal of money on layout, money that then was used to promote the booklet.

Dollars and sense

The booklet budget figures heavily in format discussions. Unusual page shapes cost more than standard shapes because part of the paper is wasted when the stock is cut.

Cost may determine the color proposed for a booklet. Four-color transparencies, and the four plates required to print them, always cost considerably more than a two- or one-color project. But perhaps that extra cost can be justified in terms of attention-getting value. If your booklet will be one of many offered, you may need color to entice potential readers.

When the budget simply won't allow two or four colors, ingenuity comes into play. Specialists in graphics and production often coordinate colored paper with colored ink. However, be sure to check the readability of your eye-grabbing paper-ink combination. Dark red on light blue might be fine for line designs, but in small print, it's hard on the eyes.

Your choice of photos or sketches depends on availability.

If no photos are obtainable and there's little time to take them, you might opt for sketches. But if no artist is available, you might decide to beg or borrow photos. Some booklets, however, are so simple and straightforward that no illustrations are needed, just attractive typography.

All these aspects must be discussed before you write the preliminary budget proposal that most companies require. In writing the proposal, getting it approved and following it, you'll be going a long way toward preventing misunderstandings, not to mention cost overruns.

Not all things in life are free

At one time, most booklets were offered free. Consumers had only to write or call and the desired booklet would be mailed. But the costs of mailing and printing have risen steadily. Now, nearly all booklets have either a postage charge or a nominal fee to cover distribution. Deciding just what the charge should be may require some projections.

The higher the number of booklets printed, the smaller the cost per booklet. Why? Because the basic production costs (typesetting, art, keylining) are the same for 1,000 copies as for 5,000. Only the paper and printing costs are more. Those planning a booklet must estimate the number of potential readers, too, and consider how to get the booklet to them.

Timeliness is another aspect to keep in mind. Before you crank out two years' stock of copies, you must decide whether a booklet of nifty new ways to tie 36-inch scarves will be just as fresh and helpful next year.

One now, more later

Finally, will your booklet be part of a series? Many large companies like to work out a standard pattern for their booklets so readers will recognize their covers or logos when the publications are displayed. For example, an international sewing machine company used the same basic cover design on its how-to-sew booklets for many years, and sewers learned to look for the newest in the series when they visited the company's stores.

Sometimes a booklet series grows year by year, but often the entire series is planned before the first booklet is written. Such was the case when a major food company launched its information center for microwave oven cookery. A distinctive typeface was designed for the folder covers. The page size was such that a folder could be two, three or four pages, depending on the subject. And each folder was printed on a different shade of paper. Publicists announcing the series listed specific topics to be covered so that microwave owners would know what to expect in the future.

Set the timer

Set up a realistic production timetable for your booklet immediately after you decide the format. Realistic is the key word. Allow enough time for the typesetter, the layout artist, the keyliner and others to do a careful job. And you will need to set aside time between writing and typesetting for the legal department to check the copy.

Have you seen the little poster on which the designer had to push the "D" in *PLAN AHEAD* down along the edge? Failure to plan ahead can jeopardize your booklet project. And if your production deadline necessitates working overtime, your organization pays a costly penalty.

In most cases, you, the writer, will be gathering information while the layout people are at work. Then, when the layout is complete, you will write the copy to fit—exactly— the space allowed.

Booklet copy must be bright and tight (see Chapter 5). *Bright copy*—that is, light and precise copy—is a must because readers may have only a casual interest in the topic or may simply want to scan the booklet to get knowledge quickly.

Tight copy is necessary because space is at a premium and not a quarter-inch can be wasted. And don't make the mistake of thinking that because you have only a few paragraphs to write that you can dash them off in no time. Writing tightly always takes longer. You have to rethink, condense, then rewrite, until your copy is cogent and concise.

Proofreading your booklet copy, as described in Chapter

21, is nearly as important as doing a competent job of writing.

Getting it into readers' hands

Once your booklet is written, you may be asked to help plan its promotion and distribution. Suggest contacting magazines that list booklets homemakers can request. If your organization handles a lot of consumer correspondence, you could insert in each answer an order blank promoting the new booklet. Other avenues for publicizing the booklet: employee magazines, stockholder publications, trade journals and membership newsletters.

When requests come in for the booklet, see to it that they receive prompt attention. This may involve setting up a procedure for filling the orders within your department or a mailing department. Booklets often can be sealed or put into envelopes at the time of printing so only the name and address need to be added before the booklets are mailed.

For further reading

Pocket Pal, A Graphic Arts Production Handbook. New York: International Paper Company, 1973.

16. Cookbooks. The seven steps from idea to distribution

BOOKS by cooks for cooks continue to be excellent publishing ventures. Popular basic cookbooks now rival the all-time sales records of such stalwarts as the Bible and dictionaries.

Although cookbook sections of stores and libraries already are crammed with volumes that are bigger, bolder and costlier than ever, new recipe collections appear every week. Each new kitchen appliance, each new culinary trend brings with it a flock of experts with their own specialties.

Cookbooks are in demand because they offer inspiration as well as instruction. Many cooks think that if they get just one wonderful recipe from a new book, it's worth the price. Others seek a new book with the purpose of recharging their culinary batteries and renewing their interest in daily cooking.

The following material on producing a cookbook is directed toward two types of food writers: the expert who will sell a manuscript, then work with the publisher to completion; and the writer/editor who is in charge of producing a manuscript for an employer.

Cookbooks take seven steps from inception to distribution: market research, complete plan, recipe selection and testing, manuscript preparation, proofing and finalizing, book production and promotion.

This discussion of the cookbook-publishing process assumes that you have a great idea—a super angle, an apt new approach—to a food subject *and* that you are certain other cooks will want your recipes and ideas enough to pay for them. As a would-be cookbook author, you have to be as zealous about your idea as an impassioned preacher and as confident in your ability as a television performer.

CHAPTER 16
Who will buy?

Informal market research usually precedes formal study of the salability of your cookbook idea. Haunt your neighborhood bookstores and libraries, study titles, tables of contents and layouts of cookbooks. Talk with booksellers. Chat with librarians. Study catalogs from publishers and lists of upcoming titles in trade magazines.

You will notice that cookbooks confined to a particular food far outnumber the how-to-cook-almost-everything books. You will notice, too, that many of the authors have "names," thanks to their affiliations with a cooking school, a restaurant or a publication. Other cookbooks are the work of entire editorial departments of national magazines or of appliance and food manufacturers.

If, after this initial research, you still believe the concept is a good one, you are ready to invest in formal market research. Companies offering this research can be found in most cities. They may simply ascertain the number of households owning the appliance your recipes feature, or they may do an in-depth study to guide you in approaching your subject. It's wise to confer with the researchers at length before hiring them.

Knowledge of the sales potential of your projected cookbook is essential to another sale: selling your idea to a publisher. While studying the cookbook market, jot down the names and addresses of publishers whose books appeal to you. Next, consult the annual *Writer's Market* for the names of people at those publishing houses who consider new manuscripts.

The time-honored method is to write to the publishers, introducing yourself and your idea and enclosing your book outline and a sample chapter. Then you wait—and hope—and wait some more.

Telephone and tell me . . .

Lately, however, writers have begun telephoning the editors to introduce themselves and their topics in person. A

COOKBOOKS

written query can be shoved aside, even lost, but a phone call is harder to ignore.

If the editor has just bought a manuscript on that topic, the writer knows it immediately and can try another publisher. If the editor seems interested, the writer fires off the outline and sample recipes immediately, following up with another call later.

If the publisher's editor does express interest, s/he will tell you what comes next. It undoubtedly will be a complete prospectus or outline, plus eight or ten sample recipes carefully chosen to illustrate the length and breadth of your book. The outline will crystallize all your thoughts on your topic, stated succinctly and organized logically. The decision to publish—or not to publish—will be based on your idea (theme) plus the samples.

If no publisher is interested in your book idea, shelve it. Few writers have successfully financed their own publishing ventures. One woman, a food processor expert, lacked capital to pay her printer, so she entered fifty-six recipes in a high-stakes recipe contest. Her prize money paid for her book; her timing was good, and so were her public relations. She later became a national demonstrator for a food processor manufacturer. Unless you are a whiz at promotion, printing your book yourself can be merely a costly ego trip.

The green light from the publisher typically includes a recommendation on the book's page size and format. The size of page and format are of prime importance because they limit the number of recipes you can include. You will want to consider many more recipes than you can possibly use so that you publish the very best.

How and by whom your cookbook will be illustrated is another important decision to be made early. The publisher usually contracts with the artist and/or art director and determines how the volume will be illustrated. As a courtesy, the writer often sees the artist's samples before s/he is hired. Because the writer/editor often suggests food subjects and serving situations for illustrations, try to establish good rapport with your artist.

While developing the outline for your cookbook, work out

a production schedule. Publishing a cookbook has been compared to having a baby—even to the pain and joy—though it often takes much longer.

Try to put some extra time into your book schedule, so that you can either take a breather or take time to reassess your efforts. It is wise to get help while preparing the manuscript, so that you, the idea person, will not use up your psychic energy too early and find yourself too frazzled to promote it after it's in print.

Many called, few chosen

Whether you are writing your own cookbook or producing one for your employer, the third major step is recipe collection, selection and testing, which often require long months of work. If you are lucky enough to have a test kitchen, you may need only to check the files for recipes that fit your food categories. After reviewing the file recipes, you can test enough new recipes to fill the gaps.

But if your book is a collection of recipes from a certain group, such as opera singers or tennis players, you first must contact them and ask them to submit favorites. This phase alone can take weeks of writing, phoning, even begging. For this type of book, a persuasive letter (see Chapter 20) is as important as a tantalizing topic.

It's inevitable: testing recipes means tasting recipes. Even if you trust your palate instinctively, line up a taste panel to share this aspect of the task. At-home cookbook authors frequently use family members, neighbors and/or colleagues to serve as tasters.

Decide ahead of time what the criteria for recipe selection will be and stick to it. For example, if yours is to be a honey cookbook, decide how much honey each recipe should use to qualify. Don't stretch the boundaries of your topic in order to slip in some of your pet recipes. Frequently named criteria for inclusion in recipe collections: originality, nutritional value, ease of preparation, taste or flavor and visual appeal.

You must also decide on a recipe style, then set it down in typewritten form for all involved to refer to. Don't confuse your reader by calling for white corn syrup in one recipe and Karo in another. Style also includes where and how such things as baking times, yields and pan sizes will be displayed.

There's a lot more to a cookbook than recipes. A stimulating introductory chapter will define the purpose of your book; it's the ideal place to discuss the ideas you used to sell your idea to the publisher. And, if there is to be a Foreword, ask an authority in your field to write it.

And you—or a writer you can persuade to help you—will write catchy titles for the chapters and thoughtful introductions to groups of recipes. These words, though few, are extremely important to the book's success, because prospective buyers scan those lines at the bookstore.

Order from chaos

Manuscript preparation, the fourth step in the sequence, often begins long before all the recipes have been tested. Soon your stacks of scribbled-on, spotted-up recipes will be transformed into neatly typed pages. Whew! The end is in sight.

An expert typist—one interested enough in food to know granulated sugar from powdered sugar—is a must at this stage. And, after you find a crack typist, look around for an eagle-eyed proofreader. Both the editor(s) and typist can read copy aloud to the proofreader during proofing, but it's mighty hard to spot your own mistakes.

As your manuscript pages begin to accumulate and you reread them, there may come a moment of truth. Seeing your recipes and recommendations in pristine black and white may make you realize that a certain part of your book may need further creative work. Perhaps the dessert chapter needs just one more extraordinary recipe, or the directions for the cookie house are too complex.

Don't hesitate to redo, rewrite or retest. Once the pages are printed, bound and shipped, it's too late to change your mind.

At the manuscript stage, you may learn that your outline was optimistic and that your agreed-upon book length will not hold as many recipes, hints, helps and menus as you envisioned. It's hard to ax the offbeat idea that you adore, but pages will only hold so much.

It's called pagination

From your final manuscript, the printer produces proofs. And from the proofs, the pages are laid out. *Pagination* is the process of dividing galleys of printed material into individual pages.

Paginating a cookbook is particularly difficult because you shouldn't split a recipe between two pages. It's not easy to read the ingredient list at the bottom of the left-hand page, then move to the top of the right-hand page for directions. And following directions that begin at the bottom of one page and end at the top of the flip side is even worse.

Your publisher may push you to break up recipes. It would save paper and allow you to shoehorn in a few more ideas. But *insist* that each page be complete in itself. If an inch or two is left open at the bottom of a page, suggest filling it with an appropriate sketch. If the spots for sketches have already been plotted (as they should be), use the space for a menu, a bit of food history, an anecdote or a variation for the featured food.

From the pasted-up pages you prepare, the publisher will make silverprints or photocopies of the press-ready pages. It's just too expensive to make changes at this stage, but you should look over the copies carefully to be sure everything jibes.

Try not to leave many details in the hands of the publisher's employees. They work on several books simultaneously and always seem to be pressed for time. One item to be especially careful about is the *running heads,* those little guidelines in the upper left- or right-hand corner of the pages. The running heads often are pasted up at the last moment. You'll be awfully disappointed if you leave the checking to someone else and discover a page of casseroles carrying a running head for desserts.

While the presses roll and the binders stitch, the progress of your cookbook is temporarily out of your hands. This period—it may be weeks or months—is the perfect time to plan the promotion.

Much as publishers and writers hate to admit it, many fine books do not reach their full audience because they were not properly promoted. Study the chapter on public relations, then map a promotion campaign on which you, your publisher and your organization can collaborate.

If you can do only one thing to publicize your new book, get the very first copies and send them to leading magazine and newspaper food editors for review. Be sure to include a personal note, mentioning your availability for an interview. These influential writers will want to assess the book while it's news.

For further reading

Evans, Nancy, and Judith Appelbaum. *How to Get Happily Published.* New York: Harper & Row, 1978.

Goulart, Frances S. "Promoting the Book, Promoting Yourself." In *How to write a cookbook—and sell it.* Port Washington, N.Y.: Ashley Books, Inc., 1980.

Hill, Mary, and Wendell Cockran. *Into Print: A Practical Guide to Writing, Illustrating and Publishing.* Los Altos, Calif.: W. Kaufmann, 1977.

17. Speeches. The writing that requires rehearsal

IF ONLY public speaking were as easy as talking one-on-one. Many of us become self-conscious, even tongue-tied, when giving speeches.

Whether you're specializing in food, families, furnishings or fashion, your speeches are "you-nique": you do the communicating. No one else can give *your* program. Your information based on your particular research; your voice, energy, enthusiasm, poise and confidence are unique.

Giving a speech or demonstration is a lot like planning and giving a party. The audience becomes your guests, participating in what you have planned. With this approach in mind, you should learn as much about your audience ahead of time as you can (review Chapter 2).

Become familiar, too, with the setting for your presentation. If possible, check the ventilation and seating arrangements in advance. If the seats are crowded or not very comfortable, you may want to plan a short break during your talk so your listeners can get up and stretch.

The content of your talk can be compared with a dinner party menu. You plan the food well in advance and spend a good bit of time seeing that the courses are balanced and that each detail has been thought out.

Your talk should have a closely defined purpose or theme (as described in Chapter 4), just as does a carefully done piece of writing. And it may be necessary to "translate" your content from language for reading to language for listening; the latter requires more repetition and simpler sentences.

As you would plan a timetable for your party, so should you consider the timing of your remarks, leaving a few minutes at the end for informal questions and comments. All public speaking takes is the ability to think while on your

SPEECHES

feet. If champion debaters and top-notch trial lawyers can do it, you can, too.

Mark it special delivery

Speakers with a great deal of information to present are sometimes criticized for putting too much emphasis on content and not enough on delivery. If you have your talk well in mind—and you must—you can establish good eye contact with your audience and quickly adjust your rate of speaking or perhaps even shorten your remarks, if your audience becomes restless.

Choosing the main points and illustrative anecdotes for your talk requires the same thought and selectivity as picking the right foods for a party. No host or hostess would serve everything s/he knows how to prepare at one dinner, nor would a speaker try to share everything s/he knows about a topic in an hour-long program.

To carry the analogy a step further, think of your opening remarks as the appetizers and your conclusion as dessert. What sort of taste will your presentation leave in the mouths of your audience?

Still another way to test a talk is: are you talking *at* your listeners, *to* them or *with* them?

The eyes have it

And what about the visual element? How will you look?

Most experts recommend dressing attractively but conservatively when appearing as a speaker. Women should not wear party clothes or too-obvious makeup. If possible, plan a brief rest before a talk or demonstration so that your levels of physical and psychic energy will be high and you can respond fully to your audience.

Will you use charts and graphs? Food and family topics lend themselves readily to the use of charts and projected images. When discussing the vitamin content of foods, for instance, you can turn to a series of charts showing groupings of foods high in each vitamin.

Movement, such as electronic waves in a microwave oven

or heat in a convection oven, can be depicted easily—and just as easily understood—with a carefully drawn diagram.

When your material involves columns of numbers, figures in the millions or statistics, you will, of course, make them visual. You should also be sure to suit them to the audience, space them and/or repeat them.

The rule of thumb is that the visual ought to reinforce—not replace—your explanation. (For information on audiovisual communication, see Chapter 19.)

Dress or not, have a rehearsal

Just as smart cooks try out a new dish before making it the *pièce de résistance* of an important occasion, smart speakers benefit from rehearsal. A run-through in front of the mirror is good; a rehearsal for a colleague or family member who will give constructive criticism is better. If your organization has a videotape machine, ask to use it. Such a tape is invaluable in helping to spot mannerisms and awkward poses.

On stage

As speaker you play a dual role: the life of the party and the host or hostess. All your planning and preparation is focused on the moment you are introduced and you step out to greet your audience.

In general, women find it easier to warm up to an audience than men do. However, despite the excitement of being the center of attention, many women are a bit uncomfortable when cast as authority figures. They tend to apologize for having the audacity to tell others what's what and to discount their knowledge or the value of their work.

Men and women who speak regularly—and enjoy it—advise first-timers to shake off such inclinations, step up with a smile, speak with enthusiasm; share your information and ideas with candor and confidence.

If you are seeking speaking experience, sign up for the *speaker's bureau*—an organization offering speakers to community organizations—of a professional group or of your

SPEECHES

business. (If there is no speaker's bureau in your community, help organize one.) These voluntary talks usually are made to small, receptive audiences, such as Parent-Teacher Association groups.

Another way to take that first step to the speaker's rostrum is to appear on a panel. Many professional and community meetings involve workshops in which several people present part of the material. Or you might serve on a reaction panel. That is, after the principal speaker gives the address, you, as a listener, share your reactions.

Whenever you speak, remember the speaker's golden rule: Speak unto others only so long as you would have them speak unto you.

For further reading

Linver, Sandy. *Speak Easy: How to Talk Your Way to the Top.* New York: Summit Books, 1978.

Stone, Janet, and Jane Bachner *Speaking Up—A Book for Every Woman Who Wants to Speak Effectively.* New York: McGraw-Hill, 1977.

18. Demonstrations.
The four types and how to organize them

DESIGNING a demonstration, then presenting it, is a demanding role.

Your design takes its cue from your purpose. Typically that purpose is to show how to do something or how something works. Hence, the nickname: *show-how*. The top-of-the-line demonstration, however, reaches a bit higher. Strong demonstrations inspire, influence and animate their audiences. They may sell a product or introduce a technique, but, in the process, the observers get a refreshing lift.

Your demonstration's design will take into account visuals, equipment, content, time and space, and, of course, your skills, voice and presence. The challenge of demonstrating is combining down-to-earth practicality and sky's-the-limit creativity.

The demonstrator's role takes both flair and follow-through to make every minute on stage—or on camera—count; to lend spontaneity to your carefully conceived presentation. You must turn your audience's passive interest into an active one. You must consider their needs in order to fulfill yours. You want to sell an appliance, to introduce a cooking technique, to make a positive impression for your employer. Your audience, however, may want to save time on food preparation, to save money on clothing alterations, to begin a rewarding hobby.

Let's look in our mind's eye at your would-be demonstration. There you stand: confident, concise, friendly, fluent. And your audience: quiet, attentive, smiling at your quips, nodding responsively.

Then the bubble bursts. You remember demonstrations

DEMONSTRATIONS

you have watched as a member of the audience. Remember? A bit bored, perhaps tired, you let your mind—and eyes—wander. Though still listening, you fidgeted, searching idly through your purse. Meanwhile, the demonstrator plunged ahead, grabbing the next accessory for the food processor. The look on the demonstrator's face said: "The hour must be filled, the plans must be carried out, regardless of response." Past experiences have taught you more about what *not* to do than what *to* do.

Types of demonstrations

Most demonstrations fall into these categories:

- Display, plus a few words of explanation, as in a television commercial or at a fair or convention
- Presentation of a home product to prospective customers, members of the press or a sales force
- Brief, dramatic demonstration to colleagues, classmates or viewers of a magazine-style television program
- Full-length class presentation

Each one of these formats has unique requirement, as we'll see in the following case studies. The theme for all four is: Sandwiches for Breakfast.

INFORMATION AT A GLANCE. The success of the display with explanation depends largely on visuals. A good visual must be simple, direct, distinct and easy to understand. A colorful and vivid display attracts attention. For example, in the background, a four-color poster could depict the *Croque Monsieur* sandwich in a French setting. A poster, incidentally, is an excellent place to hide the cue cards that all demonstrators need. A cheeseburger with fruit garnish could be displayed on a pedestal plate. And you could be baking (in a handy toaster oven) hash-topped muffin pizzas for tasting. Your commentary, or the voice-over of the commercial, mentions each sandwich. The closing offers the recipes.

One-to-one is the rule for visuals: one idea to one visual (sketch, graph, poster or motto).

Still another rule: Don't exceed 40 characters (count both letters and spaces) across a visual, whether it is a poster or a slide.

MAY I PRESENT... Your knowledge of the home product's usefulness is the key to presenting it. While preparing the breakfast sandwiches by using a new cheese sauce mix or a new table-top broiler, you explain the new product's price, availability, energy use, care and advantages.

KEEP IT SIMPLY SPECTACULAR. The short, dramatic "demo," particularly one exciting enough for television, is the most difficult to present. To do one skillfully requires detailed planning, rehearsals and self-confidence. To grab audience attention, you must start with something spectacular, talk about it, then show how to do it.

The sandwich specialist comes on carrying a sword skewered with Monte Cristos, describes the rich ham and cheese sandwich, batters it, puts one together and fries it. Then, s/he uncovers a huge tray of fruit, meat, cheese and bread for making the breakfast combinations. In closing, s/he displays a colorful accordion-style booklet showing more sandwiches and explains how to obtain the booklet.

CLASS, COME TO ORDER. Giving a demonstration-class of 50 minutes or longer demands stamina and coordination, but it is somewhat easier for the demonstrator/teacher. Why? Because the students are there because they're interested. You need not talk nonstop, because they must study their direction sheets and take notes. But timing has to be perfect, so keep one eye on the clock and the other on your outline.

The sandwiches-for-breakfast idea is a natural for a class. While demonstrating stick-to-the-ribs recipes, the teacher can weave in all sorts of helpful tips.

It looks easy—when you're organized

Whichever demonstration format you're working with, organization is the key to success. Your organizational efforts should be applied to equipment, props, visuals and script, then, those four need to be integrated.

Here are some preliminary questions to ask yourself: How much material can I present well within the time frame? Do I like this treatment/approach? Am I confident in my performance? What is the minimum amount of equipment I need? The maximum (when transporting equipment is no problem)? How many props (flowers, candles, etc.) do I want or need? (By now, you realize that the demonstrator is one-third movie star, one-third super salesperson and one-third packhorse.)

Preparing a typewritten format like the one shown here is helpful.

After you've answered the preliminary questions above, there's another set: Will everyone in the audience be able to see what I'm doing? What about the background? What shall I wear? What can I weave into my script to help my audience remember my demonstration? What take-home item can I create? Who would be a supportive—yet critical—listener for my rehearsal? Where can I get help and advice in preparing posters?

Thoughts and afterthoughts

As with so many skills, demonstrating can only become smooth and natural with practice. But we can benefit from others' experiences, too. These suggestions were collected from teachers, talk-show cooks and public utility advisers.

- Take a deep breath, smile and follow your plans. Don't be surprised if your throat gets dry and your palms get sweaty as you open your presentation.
- Move deliberately, naturally. Don't race.
- Arrange the elements of the demonstration to clarify relationships. Don't leapfrog from item to item.

TELEVISION DEMONSTRATION
(Program Segment)

Name: Leslie
Date: June 10
Subject: Sew-how with synthetic fabrics
Purpose: To point out the sewing machine adjustments necessary in sewing synthetic fabrics.
Properties: Demonstrator will bring a sewing machine and examples of synthetic fibers and blends with natural fibers. Will need a table to work on.

Time Estimate	Say	Show
1 min.	Beautiful fabrics made from synthetic fibers or blends with natural fibers now ready for spring sewing. Have some tips ready to pass along to make your sewing most successful.	Show examples of fabrics.
3 min.	Why is stitching synthetics different from natural fibers? Discuss difference between single filament and staple yarns.	Point out trouble spots on sewing machine. Use all dacron for specific example.
2 min.	Point out importance of avoiding stitching with the lengthwise grain.	CU on small sample of stitchery on synthetic fabrics.
4 min.	What adjustments are necessary on the machine for stitching? 1. top tension 2. pressure foot adjustment 3. needle hole and needle	Demonstrate making these adjustments on the machine.
2 min.	Discuss other points on the various types of synthetic fabrics.	Show other examples of fabrics.
1 min.	Close: Remember it does take these special sewing machine adjustments to sew synthetic fabrics. If you learn these adjustments you'll enjoy using these kinds of fabrics more.	Point out on sewing machine.

DEMONSTRATIONS

- Plan your transitions as carefully as your major points, so you can move easily and smoothly from step to step. Don't stall.
- Keep the commentary light, conversational. Interject a rhetorical question. Make an aside. Don't drone on.
- Pack your comments with information, offer variations, suggest other sources. Don't digress.
- Speak with confidence. Don't mutter or mumble.
- Adjust your demonstration in response to your audience. Don't cling to a plan that's producing puzzled looks or boredom.
- Leave time for questions and comments. Don't run overtime.

19. Slide talks and filmstrips. The 10-step process from idea to screen

WHEN it comes to audio-visual presentations, there's more than meets the eye and ear—more thought, more time, more work. Those who see a colorful and informative slide show never dream that it was months in the making. Not months of continuous work, to be sure, but many hours of work spread over several months; work by the writer (yes, you), other subject-matter specialists, producers, photographers, artists, models and narrators.

To meld *audio* (words) and *visual* (pictures) in your presentation, you need to follow ten steps:

1. Determination of need
2. Statement of purpose
3. Definition of audience
4. Creation of treatment/outline
5. Agreement on specifications
6. Draft of script and preparation of storyboard
7. Completion of script, including both narration and pictures for reference*
8. Evaluation and revision of script
9. Actual shooting and recording
10. Duplication and dissemination

*If a study guide is to be used, it should be prepared by the time the script is completed.

The steps in your audio-visual effort are much like the answers in a crossword puzzle: it's a lot easier to get the successive words when you've gotten the first words right.

Before we discuss these steps, we should survey the various types of audio-visual presentations and their uses.

SLIDE TALKS AND FILMSTRIPS

The audio-visual range

Audio-visual presentations run the gamut from 30-second *spots* to 30-minute motion pictures. Spots (also called *shorts*) for radio or television are a challenge because of their brevity. The writer must attract the audience's attention immediately—or not at all. You can present only one thought, so it had better be the right one.

Good examples of the demands of writing short were the "Bicentennial Minutes" of 1976 presented on television. Sixty seconds of history were dramatized for easy assimilation, and for a year a different historical "minute" was telecast each day.

Overhead transparencies are easy to prepare if you have the special equipment required, plus they offer great versatility. You can combine photographs, line illustrations, charts and graphs.

Slides and films are used widely for good reason. They lend vibrancy to many commonplace topics, yet can be prepared with the popular 35mm camera. If you want to be able to rearrange and revise your presentation, a slide show is best. If you need to distribute your presentation widely, a filmstrip is the answer. (Your presentation can be distributed with recorded narration or a narrator's script. It's up to you.)

Videotaping, an exciting newcomer, is catching on fast. Costs of portable cameras and recorders are surprisingly low.

A filmstrip—and how it grew

Let's follow the ten steps in developing a filmstrip from its beginning.

IS IT NEEDED? The state agency you work for has studied garbage disposal and has come to the alarming conclusion that trash could overtake the nation more quickly than Communist enemies could. Your boss has obtained financing for disseminating the information. Step one—*determination of need*—has been taken.

TO ALL INTENTS AND PURPOSES. With need established, it's time to frame a *statement of purpose*. You and several researchers have been assigned to develop the presentation. You get together and discuss how the film should change the actions or attitudes of the audience.

Each one *thinks* they understand the purpose: to teach others what they learned. But agreeing on a precise purpose proves difficult. You wrestle—verbally—with the words, scratching out, scribbling in. Often during this sometimes-frustrating process you can come up with a working title for the film as well as a theme. Finally, the purpose is hammered out and approved by your supervisor. The value of a well-focused and concise mission statement will be realized time and again as you proceed.

By virtue of her or his job, your supervisor is the boss of the filmstrip project. In industry, the board of directors or a department committee may be producing the strip. In that case, the writer must insist that one person speak for the group in making decisions about the strip. Inevitably, politics exist within such a group. Insist the group itself work within that framework to name a spokesperson. Naming a spokesperson will save both time and money for the group.

WHO WILL VIEW? *Defining the specific audience* comes next. A longtime scriptwriter notes that she must know "three I's" before she can begin work. They are the levels of *information, interest* and *intelligence* of the future audience.

A concise paragraph describing your audience will prove valuable as you make many decisions along the production path.

RESEARCHING YOUR TREATMENT/OUTLINE. *Several types of research must be done* at this stage. You should double-check your facts and figures, augmenting them with national or industry statistics. You should check whether photographs of your subject, or graphs and other designs already exist, and whether you can use them. You should start thinking about a photographer and a narrator. And, you should be aware of the budget range and the probable deadline for the entire project.

Before you proceed, *design the treatment/outline* (usually

SLIDE TALKS AND FILMSTRIPS

called *treatment* for short). In finished form, this is a brief document that not only describes the look, style and format of your filmstrip, but also describes in modest detail the scope and sequence of its content. A well-developed treatment focuses on the shape and appeal of your presentation. When finished, the treatment/outline will prove invaluable in estimating costs and guiding the detailed development of the script that follows.

As with any carefully conceived communication, a treatment/outline and script ought to have an introduction, the development of major points and a summary.

Of course, the middle section is the longest. That is where you play out your treatment theme and its variations. That is where you extol the features of your product or amplify the benefits of your program.

GO GET YOUR SPECS. With the treatment/outline in hand, *the specifications (specs) need to be worked out and a budget set.* That means pinning down the cost of the photographer, the slides, the narrator, the recording studio, the music and the reproduction of the finished film. Although some preliminaries can be done, such as comparing photographers' fees and checking prices of competitive photo labs, the finished specs must wait until you know how many slides, how many minutes of narration and music will be recorded, etc. Yes, this is the time to get help from a production-services professional.

STORYBOARD AND SCRIPT DRAFT. *The storyboard is your visual script.* During all the discussions of purpose, audience and treatment/outline, some members of the group have been roughing out sketches for the storyboard phase of the filmstrip. Don't worry about not being a trained artist. You can resort to stick figures or cut-and-paste figures from magazines.

While continuity is important to audience understanding, it is not necessary to picture each step chronologically. Often a visual hole or two can be plugged, thus speeding the film's pace.

The mechanics of working on the storyboard are up to you.

Index cards are used most frequently. The card can show: size and view of shot, place in sequence, instructions to photographer, plus some rough narrative. *Storyboard sheets*—a series of TV-screen shapes with lines below—work well, too. Or, you can line it all up on a legal-size tablet. Pick one method and stick with it.

As you work with the storyboard and first draft of the narration, you are faced with a massive job of organization. Writers who have done dozens of slide talks and filmstrips say their efforts may go through four or five versions. So, settle on one approach and, in the words of a veteran scriptwriter, "tough it through." That first storyboard and accompanying narrative will give you something to build on—or, something to tear apart and rebuild.

Remember, it's the sequence that counts—and the emphasis. Here is where you must develop the ability to cut through to the bare bones. Don't get hung up on details.

WORDS TO LISTEN TO. *Writing the script* for a filmstrip or slide show is a triple challenge. First, the words must supply the information that pictures cannot; second, the words must fit the length of time a scene will be on the screen; third, the words must flow easily and conversationally. There's no place for pomposity or erudition here, but there's plenty of room for natural humor and common sense.

And be sure to let the viewer know early in the script how the forthcoming information will be of help to her or him personally.

The usual script format calls for the visual portion in the left one-third of the page and the audio commentary in the right two-thirds, double- or triple-spaced. Either miniature sketches or thumbnail descriptions can be used for the picture portion. Put only four or five shots per page, so the script is easy to scan and there's plenty of room for critiques.

Pacing is extremely important in scriptwriting. Don't race through the material, but don't poke along, either. Listen for the rhythm of the words as they are read aloud or as you play back a tape recording.

Be sure you're building toward a climactic sequence, pref-

SLIDE TALKS AND FILMSTRIPS

erably near your clincher of a conclusion. In certain filmstrips, one or two of the pictures may be so dramatic, so rich in detail, that the narration can fade into the background.

"E" DAY: EVALUATION DAY. Audiovisual writers often ask typical viewers and experts in their field to *preview and evaluate* their scripts before investing in photography, narration and recording. These objective, yet friendly, outsiders often can spot potential pitfalls—skips in sequence or lapses in logic—that the deeply engrossed writer has missed.

Here's how one major food manufacturer, producer of many a well-received how-to-do-it film, handled the evaluation. Persons who were representative of the proposed audience and who were articulate and interested, whether teachers or salespersons or consumers, were invited for a preview of the how-to sequence. For the preview, staff members worked behind a framework similar to a puppet-theater stage. Each setup was coordinated with the proposed script. Evaluators were given copies of the script and note pads. Afterward, refreshments were served during an informal reaction session.

ROLL 'EM. Once the script has been approved, you can begin *shooting.* Because a professional photographer's time is costly, every effort should be made to have the setups ready and the participants on hand when filming is scheduled to start. Beginning filmmakers often don't shoot enough slides. Remember, film is far less expensive than rescheduling shooting.

As soon as the pictures are completed, the *narration can be recorded.* But first, read through your script while viewing the selected slides and be prepared to make final narrative changes. In doing a slide show, the sound—words and music, if used—goes on a cassette or filmstrip, which also carries the inaudible beeps that trigger the automatic switching of the slides. If a printed narration accompanies the filmstrip, be sure it includes phonetic pronunciations (pro-nun-SEEAY-shuns) of all unusual words.

During the recording and sound mixing, the scriptwriter

plays a key role. Off to the sound studio you go, armed with a stopwatch and your pronunciation list. Like photographers, sound engineers and narrators are highly skilled and highly paid, so try to avoid last-minute changes.

THE END... ALMOST. The final step, after producing a slide talk or filmstrip, is *duplication and dissemination.* The duplication is done by a film lab and the circulation by a film library, but the writer's job is not quite finished. Who else is better equipped to write the promotional material? There may be a catalog synopsis, introductory letters and publicity releases. And many organizations like to give the viewer a printed piece to read and use afterward as a reinforcement. That might take the form of a booklet (see Chapter 15), fact sheet, chart or poster.

For further reading

Hilliard, Robert L. *Writing for Television and Radio,* 3d ed. New York: Hastings House, 1976.

Kemp, Jerrold E. *Planning and Producing Audiovisual Materials,* 3d ed. New York: Thomas Y. Crowell, 1975.

20. Letters and newsletters. The requirements of these personal communications

THE LETTER, a one-to-one communication, and the newsletter, a one-to-many communication, have similar aims but dissimilar formats.

Personal letters, whether thank-yous to relatives or epistles to pen pals, are easy to write—usually—and quickly mailed. Our correspondents seldom care about sentence structure and punctuation.

The letters we produce as part of our work, however, are professional—not personal—and deserve care and consideration.

Whether a letter of application, a sales letter or a response to a client or consumer, your letter ought to have seven ZIPCODE qualities: Z, zest; I, information; P, promptness; C, courtesy; O, organization; D, diction; and E, empathy.

Remember the zip code

- *Zest* is something we Americans are known for. Liveliness marks our personalities; we take pride in our friendliness to newcomers and our readiness to accept ideas. Let your letters reflect your zest for your career, your product or your project. If your face-to-face conversation is breezy and filled with good humor, your written conversation can be too.

But beware of injecting too much of your ego into letters. Inspect the rough draft carefully for too many sentences starting with "I," too many personal details. And while you're counting the "I's," count the "you's" and "we's," for comparison. "You" seldom fails to hook the receiver.

- *Information*—requesting it or responding to a request—frequently is the purpose of a letter. Be sure you cover the "five W's" *(who, what, why, when* and *where)* in your letters. When you include complete information you show high consideration for the receiver.

Say good-bye to letters that are rambling, inconclusive or full of nothing in particular. Say hello to letters that are concise, definite, brief and to the point.

And phrase your letters affirmatively; but when you must be negative in tone or in fact, state the reason along with the refusal.

- *Promptness* ranks second only to information in importance to the letter receiver. How often have letters you've received included such phrases as, "as soon as possible" and, "by return mail"?

Unfortunately, letters are the last thing most of us write. If letters had hard and fast deadlines, as many articles do, they would be written more quickly. Managers of consumer-response departments often set deadlines—such as ten days—within which the response must be sent.

- *Courtesy* in letter writing goes beyond saying "please" and "thank you." The underlying attitude in your words should be consideration for the receiver.

Some beginning letter writers adopt the formal phrases used in business correspondence, such as "enclosed please find" and "please be assured." This usage is *not* a shortcut to courteous letters. These stilted, much overused phrases have been shelved in favor of natural, everyday wording. The rule remains: Write to others (friends, clients, customers, readers, the public) as you would be written to.

- *Organization* is especially important in letters because organization is the key to brevity. When you want to say your piece in one page (the most successful letters are one-pagers), you need to put first things first. What's more, you need to shorten your sentences to fewer than twenty words. Paragraphs, too, are best if kept brief: one to four lines for opening and closing paragraphs and eight to ten lines for middle paragraphs.

As with most other forms of writing, organization is accomplished by outlining beforehand and cutting afterward. But, you protest, letter writing isn't that big a deal. Why take time to outline and then to cut? Because a well-organized letter is a strong letter, and a letter with strength accomplishes what it was intended to do.

- *Diction*, your choice of words, is often overlooked in writing. In letters, your diction must convey the nuances that your voice carries when you talk—excitement, questioning, confidence, sincerity.
- *Empathy* is much easier to convey in letters when you know the receiver, of course. But try to get the *you approach* into all your letters, preferably in the first paragraph. Even when the receiver is a stranger, you should know the basic concerns of users of your product or clients for your service. For example, if a homemaker writes for help in planning a dinner party, you probably have a good idea of the type of dinners given by a typical user of your company's product.

Forms: block and simplified

The usual forms for business letter writing are the familiar block and semiblock, which most of us learned in typing class. But there's an interesting new form called, simply, *simplified*.

It is simplified because it eliminates both the salutation ("Dear Jane,") and the complimentary close ("Sincerely,"). After the complete address, the letter writer goes right ahead with the content, setting off the first short sentence with triple spacing. The first sentence frequently uses "you" or the addressee's name to engage attention. A final statement, often opening with the word "with," closes the letter; it, too, is marked by triple spacing. For example, a food writer requesting a recipe from a restaurant might close the letter, "With good food as our goal."

The simplified form for letters is especially attractive to those who write form letters to strangers. For example, publicists for food and equipment manufacturers typically open the letters that accompany their releases: "Dear Editor." Yet many of the letters do not go to the editor at all, but to a staff writer. Another typical salutation for letters to consumers is "Dear Friend." In some cases, the consumer has written to complain and is far from friendly.

By using the simplified form, these inadequate openings are avoided. Much the same situation exists with closings that are extremely formal. With ingenuity—and thought—these parts of letters can be replaced with more meaningful statements.

A newsletter for your organization

The newsletter is a popular form of communication in many organizations, regardless of size or purpose. Frequently, beginning writers get valuable experience as the newsletter editor for the local chapter of a professional group, the alumnae chapter of a collegiate organization or a community group.

A newsletter is a good starting point for several reasons. First, it is short—usually four or six pages—and does not require much time or organization. Second, you can make it a personal letter to your constituents or a mini-newspaper, complete with headlines and photos. Finally, you can get feedback from readers.

The occasional editor produces the newsletter single-handedly: writing, typing, duplicating and distributing it.

In most situations, the work will be segmented—although you remain in charge. Representatives within the organization may be *correspondents*, gathering news and writing it. You edit their contributions, but may write the lead feature yourself. You also may assign a photographer or artist to illustrate it. The layout usually falls into the editor's lap. If layout is not your forte, try to arrange to have a layout expert do it temporarily while you learn. The printing and distribution is the easiest part of the newsletter process to "farm out," because most organizations have set up systems for those tasks.

Whatever you do, keep your newsletter on schedule; when it regularly appears late, readers stop taking it seriously.

The newsletter entrepreneur

The newsletter, so versatile in form, so easy to reproduce, has proven an ideal format for specialists who want to launch a small, home-centered business. A New England cookbook writer continued her career well into retirement by publishing a one-page monthly newsletter in which she shared the economical, but elegant, recipes she gathered on her trips. A Minnesotan, after excellent experience with a milling company, saw the growing interest in microwave ovens, and quit her job to begin publishing a quarterly letter packed with recipes, tips and charts for microwave users.

If you hope to become a free-lancer, newsletter writing and editing ought to be in your repertoire.

For further reading

Feinberg, Lilian O. *Applied Business Communications.* Palo Alto, Calif.: Mayfield Publishing Company, 1981.

Monaghan, Patrick. *Writing Letters That Sell: You, Your Ideas, Products and Services.* New York: Fairchild Publications, 1968.

Wales, LaRae H. *A Practical Guide to Newsletter Editing and Design.* Ames: Iowa State University Press, 1976.

21. Copy editing and proofreading. The two final steps in writing for publication

THOUGHTFUL EDITING can always improve a piece of writing. And the key word in that sentence is thoughtful.

Gone is the image of the old-time copy editor, with a pencil behind an ear and a green eye shade. Today's copy editors may wield blue pencils or sit in front of editing terminals or text processors. Their tools are the computer keyboard, a dictionary, a stylebook, an almanac and other references and, most important, their knowledge and judgment.

First, let's clarify the difference between copy and proofs and between copy editing and proofreading. *Copy* is the written text, ready to be printed or presented (as in speeches and demonstrations). *Proofs* are text that has been set in type, ready—or nearly ready—to be printed. *Galleys, silverprints* and *page proofs* are common types of proofs.

Copy editing is correcting, amending and, if necessary, revising written text. *Proofreading,* on the other hand, is marking typographical errors ("typos"), checking spacing, adjusting copy to fit and assessing headlines. When the copy includes recipes, proofreading requires comparing the proof with the original copy to be sure that each ingredient is correct. (See copy editing and proofreading symbols in this chapter.)

Editing usually is not done by the copywriter, but by the editor of the publication or the manager of the division. Proofreading may be done by the copywriter, by a member of

COPY-EDITING SYMBOLS

	HOW THEY ARE USED	WHAT THEY MEAN	HOW TYPE IS SET
TYPE SIZE and STYLE	Lansing, mich.—	Capitalize.	LANSING, Mich.—
	College Herald	Small caps.	COLLEGE HERALD
	the Senator from Ohio	Change to lower case.	the senator from Ohio
	By Alvin Jones	Bold face.	**By Alvin Jones**
	Saturday Evening Post	Italicize.	*Saturday Evening Post*
PUNCTUATION and SPELLING	"The Spy"	Emphasize quotes.	"The Spy"
	Northwestern U	Emphasize periods.	Northwestern U.
	said "I must . . .	Emphasize comma.	said, "I must . . .
	Johnsons	Emphasize apostrophe.	Johnsons'
	picnicing	Insert letter or word.	picnicking
	theatre	Transpose letters.	theater
	Henry Cook, principal	Transpose words.	Principal Henry Cook
	days	Delete letter.	day
	judgement	Delete letter and bridge over.	judgment
	allright	Insert space.	all right
	th ose	Close up space.	those
	Geo. Brown	Spell out.	George Brown
	100 or more	Spell out.	one hundred or more
	Doctor S. E. Smith	Abbreviate.	Dr. S. E. Smith
	Six North Street	Use numerals.	6 North Street
	Marion Smythe	Spell as written.	Marion Smythe
POSITION	Madison, Wis.—	Indent for paragraph.	Madison, Wis.—
	today. Tomorrow he	New paragraph.	today. 　Tomorrow he
	considered serious. Visitors are not	No paragraph. Run in with preceding matter.	considered serious. Visitors are not
	No ¶ But he called last night and said that he	No paragraph.	But he called last night and said that he
	]Jones To Conduct[or 〈Jones To Conduct〉	Center subheads.	**Jones To Conduct**
MISCELLANEOUS	He was ~~not unmindful~~	Bridge over material omitted.	He was mindful
	one ~~student~~ came stet	Kill corrections.	one student came
	or more	Story unfinished.	
	30 or #	End of story.	———————

PROOFREADING SYMBOLS

	SYMBOL	EXPLANATION	EXAMPLE: MARGINAL MARKS	EXAMPLE: ERRORS MARKED
TYPE SIZE and STYLE	wf	Wrong font.	wf	He marked the proof.
	x	Burred or broken letter. Clean or replace.	x	He marked the proof.
	ital	Reset in italic type the matter indicated.	ital	He marked the proof.
	rom	Reset in roman (regular) type, matter indicated.	rom	He marked *the* proof.
	bf	Reset in bold face type, word or words indicated.	bf	He marked the proof.
	≡	Replace with a capital the letter indicated.	H ≡	he marked the proof.
	lc	Set in lower case type.	lc	He Marked the proof.
	sc	Use small capitals instead of the type now used.	sc	He marked the proof.
	9	Turn inverted letter indicated.	9	He ma ked the proof.
PUNCTUATION and SPELLING	∂	Take out letter, letters, or words indicated.	∂	He marked the prooff.
	#	Insert space where indicated.	#	He marked theproof.
	r ʌ	Insert letter as indicated.	r	He maked the proof.
	⊙	Insert period where indicated.	⊙	He marked the proof
	ʌ	Insert comma where indicated.	ʌ	Yes he marked the proof.
	’	Insert apostrophe where indicated.	’	Mark the boys proof.
	/=/	Insert hyphen where indicated.	/=/	It was a cureall.
	?/	Insert question mark where indicated.	?/	Who marked the proof
	em	Insert em dash, implying break in continuity or sentence structure.	em	Should we can we comply?
	n	Insert en dash, implying the word "to."	n	See pages 278 93.
	“ / ”	Enclose in quotation marks as indicated.	“ / ”	He marked it proof.
	spell out	Spell out all words marked with a circle.	spell out	He marked the 2nd proof.
	out, see copy	Used when words left out are to be set from copy and inserted as indicated.	out, see copy	He proof.
	stet . . .	Let it stand. Disregard all marks above the dots.	stet	He ~~marked~~ the proof.
	⁀	Draw the word together.	⁀	He ma rked the proof.
	tr	Transpose letters or words as indicated.	tr	He the proof marked.
	(?)	Query to author. Encircled in red.	?was	The proof read by
POSITION	¶	Start a new paragraph as indicated.	¶	reading. The boy marked
	No ¶	Should not be a separate paragraph. Run in.	No ¶	marked. The proof was read by
	=	Out of alignment. Straighten.	=	He marked the proof.
	□	Indent 1 em.	□	He marked the proof.
	□□	Indent 2 ems.	□□	He marked the proof.
	□□□	Indent 3 ems.	□□□	He marked the proof.
	eq. #	Equalize spacing.	eq. #	He marked the proof.
	⊥	Push down space which is showing up.	⊥	He marked the proof.
	[or]	Move over to the point indicated. [If to the left; if to the right]	[	He marked the proof.
	⊔	Lower to the point indicated.	⊔	He marked the proof.
	⊓	Raise to the point indicated.	⊓	He marked the proof.
	ᴗ	Less space.	ᴗ	looks better

COPY EDITING AND PROOFREADING

the clerical staff or both. Proofs should be read by the typesetting organization before they come to you or your boss.

Editing goes beyond correcting punctuation, spelling and sentence structure to revisions for accuracy, brevity and clarity (the writer's ABC's). In a few cases—the exception, rather than the rule—editing may involve rewriting all or part of the text. If extensive rewriting is necessary, the copy usually is sent back to the writer for the revision, if time allows.

Got a problem?

Here are some common writing problems that copy editors watch for:

- *Marathon sentences*: sentences that seem to run for miles. Long sentences—more than three typewritten lines—usually can be divided. Look for semicolons and/or colons as dividing places. Or try to shorten the sentence by rewording it.
- *Jargon*: "in" words that mean something to specialists but not to a general audience.
- *Nonparallel construction:* when grammatical construction ought to be parallel and isn't. Either reword to complete the parallel or recast the sentence to eliminate suggestion of parallelism.
- *Switcheroos:* switches in point of view or tense. Did the writer change from the objective "they" to the subjective "we" midway through a piece? Were both past and present tenses used in a feature or release?
- *Buried lead:* a quote or a word picture or a catchy phrase that would make a superior lead frequently can be found within the copy. Pull it up into the lead and redo the piece.
- *Overlong:* determine why the piece is too long. Is the writer's subject too broad? Could the copy be divided into one main piece and one or two shorter *sidebars* (related articles)? Was everything—pertinent and peripheral—included? Or is it just wordy?
- *Fuzzy focus:* did the writing accomplish its purpose, or did the copy go round in circles?

The college of editing knowledge

How do you teach yourself to edit? By doing it. Here are some tips:

- *Volunteer:* offer to help with editing your college newspaper or your organization's newsletter.
- *Swap editing:* find another writer who would like to learn about editing and swap stories, each editing the other's work. It's always easier to spot problems in someone else's writing than in your own.
- *Before and after:* be sure to save a copy of the material that you submit to your manager or editor. After it has been published, compare the finished version, line for line, with the original. You may be surprised at what you'll discover, and perhaps, you can profit from it.
- *Critiques:* if your writing is published with little or no editing, seek out a respected colleague and ask him or her to critique your copy, making it clear that you welcome positive *and* negative comments. In reviewing the critiques, you will learn your strengths and weaknesses and find out how to spot them yourself. But remember, no two writing efforts are alike, and no two editing efforts will be either.

The proof of the project is in the reading

While copy editing may dovetail with other tasks, proofreading is a drop-everything-and-do-it-now job. Printing production schedules never seem to have a bit of air in them, so read proofs immediately and return them the same day, if possible. But don't hurry so much that you miss errors.

Will you read proof by yourself or as half of a team? It depends on how large your staff is, what your office practice has been and how sure you are of your proofreading ability. The solo system calls for reading the proofs silently, concentrating on what the copy says and how the page (or galley) looks. Reading with the copy on your left and the proofs on your right works well for right-handed people; it's easy to compare one version with the other. The disadvantage of the solo system is that you know the material well because you wrote it, and you may overlook obvious mistakes.

In the duo system, one reader (copyholder) holds the original manuscript, reading aloud from it, the other studies the proof. This system not only introduces a fresh pair of eyes, but also brings attention to awkward constructions.

While the copy editor can—and should—revise and redo the text as necessary, the proofreader should change as little as possible. Once type has been set, every change is costly and time consuming. Squelch the urge to pick nits, making belated editing changes.

Seek and ye may find

And how do you spot mistakes that are lurking in those proofs? Read carefully while looking for certain types of errors:

- *Wrong words:* the copy says "tried," but the proof says "fried." Speedy scanning might not pick up that mistake.
- *Words that don't look right:* suspect misspelling? Look it up. Some people stumble over double letters in such words such as "attune" and "accommodate."
- *Homonyms:* don't slip on the paring—or the pairing—of the pear. Beware, too, of words that sound alike, such as "terrine"—the pan in which pâtés are baked—and "tureen," the container from which soup is served.
- *Breaks:* always double check the lines at the bottom of one column (also called *leg*) to see that the text continues correctly at the top of the next column.
- *Jumps:* never fail to check to see that the jump line (also called *reefer*—short for referral) at the end of one page gives the correct key word and page number to turn to.
- *Cross-references:* when there's a reference to a sidebar, a recipe or a chart within a copy block, be sure the page mentioned has the other material.

All the copy that fits

Frequently, the proofreader/copywriter is expected to *copy fit* (make a piece of copy fit the space) while proofreading. Whether the copy is too short or too long, count the number of lines involved and stick to it. When the copy is too long, look first for *widows*—words left alone on a line at the end of a paragraph. Eliminate the widow and save a line by cutting a word or two within the last sentence of the paragraph. Some publications have a policy of eliminating all widows. Widows or not, the ends of paragraphs are the easiest place to trim.

Copy can be lengthened in several ways. The simplest one is to break long paragraphs into shorter ones. In recipe copy, you can add a variation or a suggestion for adapting the recipe to a microwave oven or a slow cooker. Or, you might round out recipe copy with a menu featuring one or more of the recipes. In a long collection of recipes, *blurbs* (introductory comments) can be added or subtracted to make copy fit perfectly.

For further reading

Martin, Phyllis. *Word Watcher's Handbook Including a Deletionary of the Most Abused and Misused Words.* New York: David McKay, 1977.

Miller, Casey, and Kate Swift. *The Handbook of Nonsexist Writing for Writers, Editors and Speakers.* New York: Lippincott & Crowell, 1980.

Ross-Larson, Bruce. *Edit Yourself, A manual for everyone who works with words.* New York: W. W. Norton & Co., 1982.

AFTERWORD

THE FUTURE, with all its possibilities, belongs to those who prepare for it.

And what will the future hold for communicators?

- The spread of the electronic revolution.
- The expansion of specializations within occupations.
- The placing of concern for the individual and the family in a worldwide context.
- The continuation of interest in finding simplicity amidst complexity.

Those seers, the futurists, challenge us to use today to the very best advantage. They suggest:

Learn a second language—or a third.
Study computer programming or programmed learning or laser design.

And as communicators, we ought to:

Observe more closely.
Question more readily.
Think more intuitively.

Let these words from the *Sanskrit* guide you: "Today well lived makes every yesterday a dream of happiness and every tomorrow a vision of hope.

"Look well therefore to this day."

INDEX

ABCs of writing, 19-21
Accuracy, 19-20
Advertisement, creation of, 30-31
Advertising, 28-32
  audience and, 77
  government regulation of, 38-39
  recipes in, 31, 48
  yourself, 31-32
Audience, 13, 70, 94, 98, 101
  advertising and, 77
  defining, 106
  identifying, 5-7
Audio-visual presentations, 104-10

Blurbs, 122
Booklet writing, 42, 82-86
Brainstorming, 15
Brevity, 20-21

Cafeteria menus, 58-59
Clarity, 21
Columns, 60, 67, 76
Communication, 3-4
  R/C recipe for, 8-12, 41
Computer analysis of nutrients, 52-53
Consumer articles, 44
Cookbook writing, 87-93
Copy, 116
  editing, 116-22
  writing, 17-21
  writing to fit space, 9, 121
Copy fitting, 121

Demonstrations, 98-103
Diet recipes, 54
Direction writing, 41-45

Editing, 116-22

Fact sheet, 74
Family
  articles and columns, 60-66, 76-81
  audience, 5-7
  audio-visual presentations, 104-10
  booklets and folders, 82-86
  creative ideas, 13-16
  demonstrations, 98-103

how-tos, 41-45
illustrations, 34-40
newsletters, 111, 114-15
news releases, 69-75
speech writing, 94-97
Fashion
  articles and columns, 60-66, 76-81
  audience, 5-7
  audio-visual presentations, 104-10
  booklets and folders, 82-86
  creative ideas, 13-16
  demonstrations, 98-103
  how-tos, 41-45
  illustration, 10, 34-40
  newsletters, 111, 114-15
  news releases, 69-75
  speech writing, 94-97
Feature articles, 60-66, 76
Filmstrips, 104-10
Focus, 21, 119
Folders, 82-86
Food
  advertising, 30-31
  articles and columns, 60-66, 76-81
  audience, 5-7
  audio-visual presentations, 104-10
  booklets and folders, 82-86
  cookbooks, 87-93
  creative ideas, 13-16
  demonstrations, 98-103
  how-tos, 41-45
  illustrations, 34-40
  meal plans and menus, 55-59
  newsletters, 111, 114-15
  news releases, 69-75
  recipe writing, 46-54
  speech writing, 94-97
Food processor recipes, 54
Free association, 15
Free-lance writing, 76-81
Furnishings
  articles and columns, 60-66, 76-81
  audience, 5-7
  audio-visual presentations, 104-10
  booklets and folders, 82-86
  creative ideas, 13-16

INDEX

demonstrations, 98–103
how-tos, 41–45
illustration, 11, 34–40
newsletters, 111, 114–15
news releases, 69–75
speech writing, 94–97

Galleys, 116

High-altitude recipes, 54
How-to writing, 21, 41–45

Ideas, creative, 13–16
Illustrations, 33–40
 in booklets, 83–84
 to dramatize writing, 10
 five W's of, 37–38
Instructions. *See* Direction writing
Interviews, 61–63

Jump line, 121

Leads, 23–27, 79, 119
Lede, 23
Leg, 121
Letter writing, 111–14

Magazine articles, 76–81
Marketing
 cookbooks, 88
 free-lance articles, 77–78
Meal plans, 55–59
Menus, 55–59
Microwave recipes, 53

Newsletters, 111, 114–15
News releases, 69–75
Nutrition data bank, 53

Page proofs, 116
Pagination, 92
Pamphlets, 82–86
Personification, 11
Photographs, 33–40
 to dramatize writing, 10–11
 federal regulations and, 38–39
 food, 36, 39–40
 with press releases, 73–74
Piggybacking, 15
Press event, 74–75
Press kit, 74
Press release, 69–75
Promotion, 93, 110
Proofreading, 116–22
Proofs, 116
Publicist, 34, 69

Publicity release, 69–75
Public relations, 69
Puffery, 73

Querying, 77–80, 89
Question-and-answer sheet, 74

R/C recipe for effective communication, 8–12, 41
Recipes
 in advertising, 31, 48
 copyright of, 53
 diet, 54
 high-altitude, 54
 microwave, slow cooker, food processor, 53–54
 writing and adapting, 46–54
Reefer, 121
Restaurant menus, 57–58
Running heads, 92

Scenario, 11
Script writing, 108–9
Selling and persuading, 28–32
Service article, 44
Shorts, 105
Show-how, 98
Side dishes, 55
Silverprints, 116
Slide talks, 104–10
Slow cooker recipes, 53–54
Space, writing copy to fit, 9, 121–22
Specifications (specs), 107
Speech writing, 94–97
Spots, 105
Storyboard, 107–8

Transitions, 21
Treatment, 106–7

Visuals, 33–40, 99–100

Widow, 122
Writers, qualities of successful, 3–4
Writing
 ABCs of, 19–21
 common problems of, 119
 dramatization of, 10–11
 focus, 21
 free-lance, 76–81
 ideas for, 13–16
 outlining, 9–10
 persuasive, 28–32
 R/C recipe for effective, 8–12, 41
 resources, 17–18
 training for, 18–19